THE OPEN UNIVERSITY

Science: A Second Level Course

Genes and Development

UNIT 1
Development: The Component Processes

UNITS 2 and 3
Cell Differentiation

Prepared by an Open University Course Team

THE OPEN UNIVERSITY PRESS

THE S2-5 COURSE TEAM

Chairman and General Editor
N. R. Cohen

Unit Authors
N. R. Cohen
J. Hambley
D. R. Newth*
J. N. Thomas

* External Consultant

Editor
Jacqueline Stewart

Other Members
R. M. Holmes
S. W. Hurry
R. Jones (*BBC*)
M. MacDonald Ross (*IET*)
Barbara Pearce (*Course Assistant*)
L. C. Pearce
S. P. R. Rose
J. Stevenson (*BBC*)

The Open University Press
Walton Hall, Milton Keynes.

First published 1973
Copyright © 1973 The Open University

Designed by the Media Development Group of the Open University.

Printed in Great Britain by Martin Cadbury Printing Group, a Division of Santype International.

ISBN 0 335 02400 9

This text forms part of an Open University Second Level Course. The complete list of Units in the Course is given at the end of this text.

For general availability of this text and supporting material, please write to the Director of Marketing, The Open University, P.O. Box 81, Walton Hall, Milton Keynes, MK7 6AA.

Further information on Open University courses may be obtained from the Admissions Office, The Open University, P.O. Box 48, Walton Hall, Milton Keynes, MK7 6AA.

1.1

Unit 1   Development: The Component Processes

Contents

Table A

List of scientific terms, concepts and principles used in Unit 1

Introduced in a previous Unit	Course Unit No.	Developed in this Unit	Page No.	Developed in a later Unit	Unit No.
	S100*	albino	11		S2-5
amino acid	10	allele	10	blastocoel	5
axon	18	aneuploidy	11	ectoderm	5
carbohydrate	10	animal pole	25	embryonic axis	5
cell	14	antibody	14	endoderm	5
centrifugation	14	autosome	10	nuclear transfer	2
chromosome	17	blastomeres	16	organizer	5
clone	19	blastula	16		
cornea	2	centriole	15		
cytoplasm	14	cleavage	16		
diploid	19	commitment	21		
DNA	17	congenital	13		
dominant	19	cytodifferentiation	16		
evolution	21	cytogenetics	12		
gamete	19	determination	22		
gene	17	egg cortex	18		
gene replication	17	embryogenesis	12		
genetics	17	embryonic induction	22		
genotype	17	gametogenesis	15		
haemoglobin	19	gastrulation	16		
haploid	19	germ-line	12		
heterozygous	19	heterogametic	10		
homeostasis	18	homogametic	10		
homologous chromosomes	18	inductor	22		
homozygous	19	in vitro	18		
hormone	18	monospermy	15		
interphase	17(TV)	monozygotic twins	10		
iris	2	morphogenetic movement	16		
isotope	6	oogenesis	15		
Krebs cycle	15	parthenogenesis	15		
meiosis	19	polyspermy	15		
membrane	14	primary spermatocyte	15		
metabolism	15	regulation	22		
mitochondrion	14	rhesus blood groups	13		
mitosis	17	somatic cell	11		
molecule	6	somatic mutation	12		
mRNA	17	spermatogenesis	15		
mutation	19	teratogen	13		
neuron	18	vegetal pole	25		
nucleic acid	17				
nucleotide base sequence	17				
organelle	14				
physiology	18				
polyploidy	19				
protein	13				
purine	17				
pyrimidine	17				
radioactive	6				
recessive	19				
red blood cell	14				
ribosome	17				
sex chromosome	19				
transcription	17				
translation	17				
unicellular organism	18				
X-rays	28				
zygote	17				

Table A

List of scientific terms, concepts and principles used in Unit 1

Introduced in a previous Unit	Course Unit No.	Developed in this Unit	Page No.	Developed in a later Unit	Unit No.
	**S22- **				
alga	2				
cell differentiation	7				
chordate	1				
coelom	1				
epidermis	2				
fertilization	8				
gametophyte	2				
gill slits	6				
mesoderm	1				
placenta	8				
spermatozoa	8				
tissue culture	7				

* The Open University (1971) S100 *Science: A Foundation Course*, The Open University Press.

** The Open University (1972) S22- *Comparative Physiology*, The Open University Press.

Objectives for Unit 1

At the end of this Unit you should be able to:

1 Understand the terms, concepts and principles in Table A. Demonstrate this understanding by:
(a) distinguishing between the terms, concepts and principles.
(b) distinguishing between true and false statements about them.
(ITQs 1 and 3; SAQs 1, 2, 3 and 4)

2 Evaluate evidence attributing a role to genetic information (genotype) in development.
(ITQs 1 and 3; SAQs 1, 2, 3 and 4)

3 Give, with appropriate examples, three areas of everyday life where research in developmental biology has particular relevance.

4 Name and briefly define three central processes that occur during the development of animals and plants.

5 Name or choose from a given list those things that change during development and those that remain constant.
(SAQ 4)

6 Describe or identify from given data those features that are characteristic of (a) cleavage, (b) gastrulation.
(SAQ 3)

7 Evaluate evidence attributing a role to environmental factors in development.
(ITQ 2)

8 Name and briefly describe three techniques or methods of study relevant to developmental biology.

Study guide to Unit 1

This Unit, which comprises the text plus the TV programme, is meant as a general introduction to the Course as a whole. The Unit deals with three main topics: the genetic background to development; the scientific and sociological reasons for studying development; and the component processes of development. These processes are in fact dealt with in detail in later Units of the Course. They are often highly visual. You therefore need to watch things happen and this of course means watching TV and doing home experiments. Indeed you are advised *to watch the first TV programme before studying the text of this Unit*, as it is meant to present you with a broad view of some of the problems crucial to developmental biology to be tackled in this Course.

As it is assumed you are familiar with S100, you are advised that, after watching TV programme 1 but before reading the text, you should test your knowledge of the relevant portions of S100 by doing the pre-Unit assessment test given on page 8. Then check your answers against those given on page 27 and do the appropriate re-reading of S100 as indicated, before studying the text of this Unit.

In this Unit we have used a new format. The text is printed in two columns. The left-hand column represents the bulk of the Unit. The right-hand column contains various sidenotes. These include further details of important points made in the left-hand column, asides, definitions of terms, in-text questions (ITQs) and the figures. All the sidenotes are identified by numbers corresponding to those in square brackets in the left-hand column. *It is of course intended that you study both columns*. Initially you might find it useful to read relatively rapidly only the left-hand column, so that you appreciate the main topics we deal with. Then you could study the Unit again, reading both the left-hand column and the sidenotes. A final read through the left-hand column may help in revising the Unit as a whole. How you actually employ this layout is of course up to you. We would however be very grateful to hear from you about whether it serves any useful purpose or not, or indeed if it is more difficult to study this way.

You will find self-assessment questions (SAQs) at the end of some Sections. We think that you will best benefit from these if you do them (and check your answers) as they arise. This is also true for the ITQs upon which some of the later arguments depend.

You are not required to remember all the details or names of the developmental changes described nor to commit to memory the complex diagrams shown. You should however understand the broad principles involved and some terms and concepts. Reference to Table A and the objectives should help clarify this. For example, you should understand terms such as *commitment*, *determination* and *induction* but need not at this stage worry about the details of these and other processes which will be dealt with in depth in later Units.

There is no set book for this Course. We do hope, however, that you will have time for some background reading for this and the later Units, though this is purely optional. A list of books suitable to this and the other Units is given in the *Introduction and Guide to the Course*.

Pre-Unit assessment test

You should now attempt this test before proceeding with the text of this Unit.

Questions 1–8
Which of the following statements are true and which are false?

1   Mitotic division of cells results in a reduction by half in the number of chromosomes per cell.

2   All the members of a clone of cells have the same ancestry.

3   In a complex multicellular organism several different types of cell exist.

4   An organism that is heterozygous for a particular gene has received similar copies of that gene from each of its parents.

5   A zygote arises from the division of an egg to give two cells.

6   Mutation can arise if part of the sequence of nucleotide bases in the DNA is altered.

7   There are differences in the number of chromosomes per cell between a male and female human.

8   The phenotype of an organism is entirely governed by the structure of its DNA.

Question 9
Which of the following are likely to cause mutation?
(a)   X-rays   (b)   atomic radiation
(c) γ-rays.

Question 10
Would you expect, for any particular organism, a gamete to have the same number, less, or more chromosomes than a zygote?

Now check your answers against those given on page 27.

1.1   Introduction

All adult multicellular organisms, whether they spring from egg, bud or seed, are the products of a developmental history during which the shape, size, anatomy, physiology and chemistry of the individual undergo profound changes. The nature of these changes and the ways in which they are controlled are the primary concern of developmental biologists. [1]

Clearly there must be as many kinds of developmental history as there are kinds of organism. Fortunately these developmental histories have much in common, and it is to the fundamental common factors in development, to the problems they pose and to the way these problems are tackled that this Course will introduce you.

However, the scientific study of development relies upon the fact that any two animals or plants of the same species will develop by the same processes and obey the same laws. Clearly what members of the same species have in common, and the features in which they may differ, are crucial to experimental work. We must therefore begin by outlining some of the genetic background to development (Section 1.2). This is followed by a consideration of the scientific and social relevance of studies on developmental biology (Section 1.3). The rest of the Unit is devoted to a brief consideration of the component processes of development and some guidelines as to how they are studied.

1 Developmental biology is also concerned with related processes, notably:
(a)   the life cycles of unicellular organisms
(b)   the regeneration of lost parts in animals and plants (i.e. in some animals and plants certain tissues removed or damaged may be replaced by growth of new tissue).

1.2   The genetic background

It is a matter of common experience that 'like begets like'. This is generally true within a species so that, with rare exceptions, parents and offspring belong to the same species. [2] It is also true that offspring resemble their parents more closely than they do more distantly related members of their own species. As, however, in populations where the production of offspring depends on the fusion of male and female gametes, the parents must differ from each other, if only in sex, the resemblance cannot approach identity to both parents. In practice we find that the phenotype [3] of an organism, if compared with the phenotype of its parents shows: [3a]

(a)   some features in which it is indistinguishable from either parent;

(b)   some features in which it is indistinguishable from one, but differs from the other;

(c)   some features in which it differs from both.

It is, in principle, possible that both inheritance (the resemblance) and variation (the differences) could be due to similarities and differences in the life experience of the individuals. The environment could determine what sort of organism the individual becomes. This view in its extreme form is quite untenable, though its rejection does not mean that our phenotypes are independent of the environment. As we shall see, some properties of organisms are almost wholly insensitive to environmental influence, others are more or less sensitive.

2 Two points must qualify this generalization. First, if it were always true no new species could arise. This apparent paradox comes from forcing a discontinuous scheme of classifying organisms on populations which do change with time. Secondly, new species *can*, especially in plants, arise in one generation and, by the artificial production of polyploid hybrids (see note 5, p. 11), they can be manufactured.

3 As you know from Unit 17 of S100, the visible characteristics of an organism are collectively known as its phenotype. The word is often used in a broader sense to include all anatomical, physiological and many chemical features that can be determined by more sophisticated methods than just looking.

As an example we may consider human monozygotic twins which have as much in common as it is easily possible to have. A pair of monozygotic twins are clonally [4] derived from a single fertilized egg and have shared the same uterine environment; both will often spend their childhood in the same general physical environment. Yet, though their resemblances can be striking and in many chemical features they will be indistinguishable, they do differ visibly, sometimes in quite dramatic ways.

Sexual or bi-parental inheritance endows the fertilized egg with two broadly equivalent sets of chromosomes, one set from each parent. Chromosomes can often be distinguished from one another by size and shape and it is found that each set can be matched against the other in homologous pairs. The single set is called a haploid set, the double set a diploid set. [5] During mitotic nuclear division each chromosome divides into two apparently identical daughter chromosomes, of which each daughter cell receives one. All the cells of an organism thus contain a replica of all the chromosomes contributed by the parents.

The equivalence of the haploid sets contributed by the parents is subject to two kinds of qualification. The first concerns the so-called *sex chromosomes*. It is usual for one sex to be characterized by the possession of two X chromosomes, the other by one X and one Y. The former, in producing by means of meiosis (S100, Unit 19) haploid sets for its gametes, must give each set one X and is therefore called homogametic; the latter gives half its gametes an X and half a Y and is therefore called heterogametic. Those chromosomes which are *not* sex chromosomes are known as *autosomes*.

The second qualification emerges from a detailed study of the actual genetic information in homologous chromosomes. This takes the form of serially arranged lengths of DNA, or genes, which lie end to end along the chromosome. We might expect homologous chromosomes to carry homologous genes, and in general they do. However, the actual genetic information carried by the gene may differ between the two homologues because of differences in the sequence of nucleotide bases; we then say that two alleles (or alternative forms of the gene) [6] are present in the cell. Where the two genes are identical, or at least indistinguishable, the cell and the whole organism of which it is part is said to be homozygous for that gene. Where the two alleles are different we say the organism is heterozygous for that gene.

In the production of gametes, nuclear divisions, called meiosis, lead to the formation of haploid nuclei. The haploid sets provided for the gametes are, however, randomly selected from chromosomes of paternal and maternal origin. Furthermore, during meiosis a process of exchange of genetic material between homologous chromosomes may occur, so that any chromosome in the gamete is itself a mixture of maternally and paternally derived pieces. The consequence is that each gamete will have one representative of each gene, but that overall these will be chosen at random from those ultimately derived from the organism's parents (S100, Unit 19).

The sum of the genes present in an organism will thus consist of many homozygous pairs of alleles and many heterozygous ones. Where a gene is represented by a pair of different alleles, they are often both effective in contributing to the phenotype. An example would be the ABO

3a **Table 1 Resemblance between parents and offspring**

Individuals	Sex[a]	Height[b] in m at age 25	Blood group (ABO)	Haemoglobin
Parents				
Mr LKJ	Male (♂)	1.90	B	HbA
Mrs LKJ	Female (♀)	1.82	O	HbA
Offspring				
Mrs MO (née J)	♀	1.76	B	HbA
Mr DJ	♂	1.92	B	HbA
Mr KJ	♂	1.89	O	HbA

Individuals	Blood[c] pressure at age 40	I Q[d] at age 10	PTC[e] threshold mg/L.	Blood[f] group (Rh)	Blood[g] clotting time (minutes)
Parents					
Mr LKJ	165/100	116	0.01	Rh⁻	7.0
Mrs LKJ	145/90	105	140	Rh⁻	8.2
Offspring					
Mrs MO (née J)	145/85	123	0.02	Rh⁻	7.3
Mr DJ	140/90	108	280	Rh⁻	6.9
Mr KJ	150/95	110	0.02	Rh⁻	810

(a)   Less than 1 per cent of adult human subjects are biologically ambiguous as to sex.

(b)   Height is sensitive to the experience of the individual in nutritional and other respects, but with 'optimum' nutrition offspring *tend* to lie between the mean parental height and the population mean, but do not always do so.

(c)   Blood pressure varies in any one individual from time to time. The figures given here should therefore be treated with caution. Mr LKJ is a little on the high side for a male of this age.

(d)   IQ if measured in a standard way in a particular society correlates with social and educational performance quite well. It therefore means something, but it is very hard to know exactly what. 100 is the mean score of a large population. Offspring tend to lie between 100 and the mean of their parents' score.

(e)   The minimum concentration of PTC (phenylthiocarbamide) that can be tasted varies widely in human populations. 'Non-tasters', i.e. those who cannot detect low concentrations, are homozygous for a recessive allele.

(f)   Both parents are homozygous; the children follow suit.

(g)   Blood clotting times can be measured in different ways and the time depends both upon the method and the care with which it is used. However, Mr KJ is wildly beyond the normal range and is, in fact, a haemophiliac. (Haemophilia is a disease in which the blood fails to clot.)

4 All the cellular descendants of a single cell are members of one clone, if all the nuclear divisions producing them are mitotic (S100, Unit 19).

10

blood group system. An individual, heterozygous in possessing both A and B alleles, shows both A and B substances on its cells. In other cases one member of a heterozygous pair may fail to make its presence felt. It is then designated as *recessive* to its *dominant* partner. For example, certain human beings lack pigments in the skin, a condition known as *albinism*. Most albinos are the children of non-albino parents, a situation that can be explained as follows:

The father has one 'albino' allele (a) plus one 'non-albino' allele (+). It follows that half the gametes he produces will carry the albino allele.

The mother is in the same position.

If the offspring of such a union are predicted we should expect one-quarter to have two 'non-albino' alleles (+/+) and to be non-albino in appearance, one-half to have one albino allele only (a/+ like their parents) and to be non-albino in appearance and one-quarter to have two albino alleles (a/a) and to be albino in appearance. This can be represented diagrammatically [7] as:

a/+	×	a/+		— parents
a	+	a	+	— gametes
a/a	a/+	a/+	+/+	— offspring

Such inheritance is called *Mendelian* after Gregor Mendel, its discoverer. The Mendelian calculus allows for the prediction of the genotypes and phenotypes of organisms in respect of any number of separate genes.

Mendelian inheritance implies very considerable accuracy in the process of replicating genes during cell proliferation. Yet altered alleles do occur 'spontaneously' rather rarely. They are known as gene-mutations and can take the form of change in part of the DNA sequence constituting an allele, that is, change in the nucleotide base sequence, by loss or substitution. [8] It is characteristic of mutant alleles that their own subsequent replication is also very accurate. Mutations in cells that are destined to give rise to tissues of the body (so-called *somatic* cells) are known as somatic mutations. They will often pass unnoticed. Occasionally the descendants of a somatic cell which has mutated are seen as patches of aberrant skin with different colour or hair form from the rest of the body. [9]

Modern genetics has thus provided an understanding of what lies behind both inheritance and variation in animals and plants. Because DNA contains, in the sequence of the nucleotide bases of which it is largely composed, information which specifies the amino acid sequence in protein, cells with the same complement of DNA molecules have the same repertoire of possible proteins that they could synthesize. Because the DNA molecule can be copied with great precision to produce two 'daughter' molecules with the same base sequence as the 'parent' molecule, any clone of cells (mutation ignored) will have the same genetic information and hence the same range of proteins available to all its members.

This limits the effect of the environment of the cell to controlling or selecting which part of the genetic information present in it should be used (transcribed and trans-

5 The haploid number, that is the number of chromosomes per haploid (gamete) cell, for a species is often given the abbreviation n. Most known diploid (2n) numbers are below 30, but higher numbers are found. The numbers for some favourite genetical organisms are:

fruit-fly (*Drosophila melanogaster*)	—	2n = 8
man	—	2n = 46
mouse	—	2n = 40
maize	—	2n = 20

Two interesting types of chromosomal anomaly may occur spontaneously or be provoked experimentally. Polyploidy is a condition in which a cell has more than two complete haploid sets, such as in triploid (3n) and tetraploid (4n) eggs which may, in some species, develop into adults. Within the mammalian body some tissues (notably liver) have a proportion of polyploid cells.

Aneuploidy is a condition in which the cell does not have a multiple of haploid sets, but has one, or a few, too many or too few chromosomes. Aneuploidy is often lethal, but organisms exhibiting minor degrees of it may be viable.

6 The word gene is used in at least two different ways, and sometimes ambiguously. Often, however, the context makes it clear which of the following meanings is in the user's mind. It is used:
(a) collectively for all the possible alleles that could occupy one particular site (or locus) on a chromosome.
(b) for one specified allele.

7 × is common genetic shorthand for a 'cross', that is, a mating between the individuals indicated.

8 Spontaneous mutation rates in man are estimated, for different genes, to lie between 1 in 10 000 and 1 in 100 000 per allele per gamete.

As you know from Unit 19 of S100, mutation rates can be raised by high temperature, some chemical agents and by ionizing radiation (e.g. X-rays or the radiation from decaying radioactive isotopes). Not all mutations are necessarily harmful, but on balance it is possible to say that in human populations mutagenic agents (i.e. things causing mutations) are hazardous to future generations. Hence great care is taken to minimize the exposure of the gonads (including those of a foetus in the uterus) to X-rays. Nuclear weapon fall-out is, fortunately, a diminishing hazard as overground tests have been largely abandoned.

It is important to stress that the effects of mutation on the organism are not specifically related to the agents provoking them. For example, mutations caused by X-irradiation do not necessarily confer greater or lesser sensitivity to X-rays on the organisms carrying them.

9 Somatic mutation is thought to be responsible for some cases in man in which one eye differs from the other in colour or in which one segment of one iris is different in colour from the rest.

ITQ 1

Will the stage of development at which a somatic mutation occurs affect its final effect?

Now check your answer against that on page 27.

lated) at any one time. In development, however, such selection presents a critical problem. [10]

Thus, if we consider the incentives for studying development, an important place must be given to the need to understand how genetic information is put to use in different ways in the different cells of the organism.

SAQs for Section 1.2

SAQ 1 (Objectives 1 and 2)

A human ovum containing one X chromosome is fertilized by a sperm containing one Y chromosome. Will the resultant child be male or female?

SAQ 2 (Objectives 1 and 2)

A red female flower is fertilized by pollen from a red flower taken from another plant of the same species. Some of the seeds thus produced give rise to some plants with white flowers, some to plants with red flowers. Ruling out mutation, how could this arise in a simple way?

Now check your answers against those given on page 28.

1.3 Incentives for study

1.3.1 Completing the cytogenetic scheme

The 50 years between 1860 and 1910 were a golden age for biology. Old misconceptions disappeared as evolutionary theory was vindicated and as belief in 'spontaneous generation' finally disappeared. Above all, cell theory approached its present position when the fundamental nature of the mitotic and meiotic cycles and of fertilization became clear. When the first links between genetics and chromosome behaviour were established, giving rise to the notion that genes were within chromosomes, a coherent biology of the multicellular organism was in the making. [11] But over it all hung an apparent paradox, which is only now yielding to direct investigation.

In its simplest form cytogenetics (the combined study of 'genetics and cell structure') sees a higher plant or animal as a clone of cells descended from the fertilized egg through mitotic nuclear divisions. Each cell contains the same genetic information as the others and the same as did the zygote at fertilization. Although important exceptions [12] to this general statement are known, what matters is that it is broadly true in many cases. Yet the adult organism consists of cells that are blatantly different from each other in size, shape, physiological activity and chemical constitution. The differences are essential to the varied contributions which cells make to the economy of the whole organism (S100, Units 14 and 18).

Thus the cytogenetic scheme cannot be complete until the gap between the genotype created at fertilization and the phenotype visible in later life has been bridged by an understanding of the development of the organism, a development which involves the creation and maintenance of significant differences between cells of identical genetic origin. This challenge has been recognized for many years and many approaches to it have been suggested. As we shall see, the era of molecular genetics has given a new impetus to work in this field, but we still seek general solutions to the problems of development.

10 One recently discovered model is that provided by the Lyon hypothesis—now generally accepted. This hypothesis suggests that in female mammals, whose sex-chromosome constitution is XX, one X-chromosome in each cell is inactivated (and may bear witness to its condition in some species by appearing as a visible structure, called a Barr body, in the interphase nucleus). X-inactivation takes place early in embryogenesis (the development of the embryo) and, at this time, which X-chromosome maternally or paternally derived, is inactivated in each cell is a matter of chance. But the descendants of any cell with an inactivated X remain committed to the inactivation of the same X. Here one is dealing with inactivation of a whole chromosome which bears many genes. How far the Lyon phenomenon is a model for *selective gene inactivation* in autosomes is still in question.

11 It is humbling to realize how many of the problems of modern biology were foreseen in this period. Molecular biologists, for example, are conscious of the fact that nucleic acid was first isolated by Miescher in 1871 (and christened 'nuclein') and that it was proposed as a candidate for the hereditary material by Wilson in 1895. Perhaps only one of the major insights achieved during this period sprang from the work of a single man—and paradoxically Mendel's work, published in 1866–67, remained unremarked for 35 years until it was simultaneously and independently resurrected by a German, an Austrian and a Dutch worker. The other great advances in understanding were assembled piecemeal by many workers.

12 For example:
(a) Somatic mutation can result in sub-clones of cells with different properties from the rest.
(b) In some animals the somatic cells differ visibly from the cells from which they are derived during development (the so-called *germ-line* cells) in respect of their chromosomal constitution.

1.3.2 Congenital defect

No one knows with any certainty the prevalence of pre-natal death in man. The existence of any embryo dying before the 'first missed period' will not be suspected and deaths before 4 weeks may often go unnoticed by the mother. The balance of opinion is that more than 30 per cent of all fertilized human eggs fail to come to normal birth. **[13]** The true figure could be much higher. It is, perhaps, not surprising that of those that do come to live birth, about 4 per cent suffer from some, often very slight but in some cases severe, *congenital* defect. **[14]** This may be considered to be the tip of the iceberg of prenatal disease. But in terms of human suffering it is a pretty serious tip.

The congenital defect that we see is of many kinds—failure to achieve normal anatomical structure, chemical deficiency, physiological accident. If we wish to avert it we must clearly know something of its causes. **[15]** Is it wholly genetic? Does it arise wholly from the conditions of pregnancy? Or are the diseases the result of a complex of genetic and environmental influences? There are many clues from which to judge. Let us take some examples.

(a) Soon after thalidomide was introduced as a popular tranquilliser it became apparent that many women who had taken it in early pregnancy later gave birth to deformed children. Thalidomide was thus suspected of 'causing' deformity. **[16]** Before the matter was cleared up an alternative hypothesis was presented. This was that thalidomide, far from causing deformity, rescued from an otherwise certain death embryos that were deformed for other reasons. Two arguments led to the conviction of thalidomide as a teratogen, (a deformity-producing agent). A careful retrospective study of the available evidence suggested that *nearly all mothers* who had taken the drug early in pregnancy produced affected children. It seemed impossible that nearly all embryos were doomed to have the kind of deformities associated with thalidomide. Secondly, direct experiment on the embryos of some other animals showed unequivocally that thalidomide damaged them in the same way. It is therefore reasonable to view thalidomide as a powerful poison for human embryos, which damages embryos of many genotypes. Of course, the genotype of the embryo, or of the mother, may affect the severity or the kind of damage to some extent, but in practice thalidomide alone is treated as the culprit. **[17]**

(b) A number of other drugs are associated with infant misery of a less severe and less constant kind. The smoking of tobacco or the drinking of alcohol by the pregnant mother increases the risk of a sick infant. The risk, however, does not approach 100 per cent even for heavy smokers or drinkers. It is thus possible to believe that whether or not an infant suffers in these cases is partly a matter of its mother's, or its own, genetically determined susceptibility to the poisons concerned.

(c) Some diseases of the new-born are products of an interaction between mother and foetus which has a known genetic basis. For example, as well as the ABO blood groups in humans there are several others including the rhesus (Rh) group. Those individuals with the rhesus substance on their red blood cells are called rhesus positive (Rh$^+$), those without are Rh$^-$. Since the Rh$^+$ genotype is dominant, an Rh$^+$ individual can arise from one Rh$^-$ and

13 There is evidence that pre-natal mortality, like post-natal mortality, falls more heavily on males. If this is so the primary sex-ratio, that is, the sex-ratio at fertilization, may be widely different from 1:1. Many workers think that about 160 male: 100 female is a reasonable estimate. If they are right, this is difficult to reconcile with a male meiosis which produces X-bearing and Y-bearing spermatozoa in equal numbers.

14 *Congenital* only means 'present at birth'. It does not imply anything about the origin of the defect.

15 It is worth thinking hard about the use of the word *cause*. In biology it may be used in rather different senses. After reading this Section ask yourself in what respects the statement 'thalidomide causes deformity' is satisfactory, and in what respects inadequate.

16

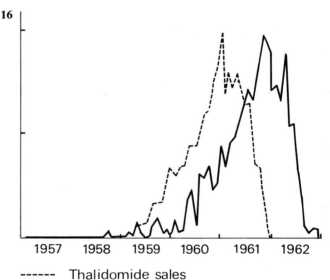

------ Thalidomide sales
——— abnormalities of the Thalidomide type

Figure 1 Graph showing the relationship between malformations of the thalidomide type and the sales of thalidomide.

17 However, thalidomide administered to pregnant rats causes little foetal damage. This could be because:
(a) rat cells are less susceptible;
(b) the rat placenta does not pass thalidomide to the foetus as readily as does the placenta of monkeys, men or rabbits;
(c) rats excrete thalidomide exceptionally rapidly;
(d) rats effectively metabolize thalidomide into harmless substances.

ITQ 2

Can you think of other *possible* explanations for the difference between man and rat in their response to thalidomide? How would you set about finding out which explanation(s) to accept?

Now check your answer against that given on page 27.

one Rh⁺ parent, or from two Rh⁺ parents. If a Rh⁺ foetus is in the uterus of a Rh⁻ mother, complications can arise if foetal cells cross the placental barrier and enter the maternal blood-stream, as the presence of Rh⁺ *antigens* from the foetus in the bloodstream of the mother can cause the mother to produce *antibodies* [18] against the Rh⁺ factor. Such antibodies can cross back into the foetus and then attach to and damage its own red blood cells. When such a rhesus-positive foetus is 'attacked' by antibodies produced in a rhesus-negative mother, the limitations of the theory of a simple genetic-versus-environmental cause of disease are readily seen. Here the mother is clearly part of the environment of the foetus, yet the attack will occur only if the genotype of both is appropriate, and indeed is most unlikely to be serious unless the mother has had previous Rh⁺ children, as only then will the mother produce much anti-Rh⁺ antibody in response to secondary contact with the Rh⁺ antigen.

But although a particular defect may, or may not, be attributable mainly to genetic or environmental 'causes', there are ways of assessing the overall genetic and environmental contributions to congenital defect. Calculation suggests a roughly equal division of responsibility between genotype and environmental causes.

Thus, the hope of controlling human suffering gives us an added incentive in our approach to development and genetics. [19] This incentive is growing as the conquest of infectious disease leaves the residuum of congenital defect, psychiatric disorder, and the degenerative diseases of old age looming proportionately ever larger in the medical burdens which society has to bear.

(d) A number of malformations and a number of biochemical defects are known to be 'inherited', quite strictly in a Mendelian fashion. When their expression is uniform in all children with the appropriate genotype, we can suggest that the disease is 'caused' genetically. Even so it is wrong to ignore the environment. As those of you who have studied the Biochemistry Course, S2-1* will know, the expression of the very unpleasant, genetically determined disease, phenylketonuria, [20] can be modified by feeding affected children from as soon after birth as possible on a diet in which phenylalanine has been severely restricted.

1.3.3 Ageing

The processes of senescence are of obvious interest to human societies. We are so familiar with them that we may forget that they demand explanation. It is not self-evident that a clone of cells should become increasingly prone to death with time. The failures of the homeostatic mechanisms of the body which become apparent with time are as much part of its developmental history as was the establishment of these mechanisms early in life.

1.3.4 Cancer

Cancers consist of cells which break some of the rules which other cells obey: they divide in an uncontrolled fashion; they invade healthy tissues; they may move and colonize parts of the body far from their site of origin. Embryonic cells do such things, but in a controlled way. It is often suggested that our investigations of embryonic control systems will throw light on the behaviour of cancer cells.

* The Open University (1972) S2-1 *Biochemistry*, The Open University Press.

18 Antibodies are proteins produced by many vertebrates in response to some foreign molecules (i.e. molecules unlike those normally found in that individual) called *antigens*, when present in the bloodstream. Each type of antibody is specific for a particular type of foreign molecule and binds to that type of molecule. First contact with a foreign molecule leads to some antibody production. Secondary contact with more of the same type of foreign molecule leads to production of a larger amount of antibody against that molecule. Antibodies help provide the animal with a defence against harmful (pathogenic) organisms such as viruses and bacteria.

19 For example, in recent years it has proved possible to detect the presence of a Rh⁺ foetus while still *in utero* in a Rh⁻ mother and to change completely the blood of the unborn infant, thus saving it from probable death. As well as advances in transfusion and obstetric techniques, this innovation also of course depended on an understanding of the Rh blood groups and their genetic inter-relationships.

20 Phenylketonuria is a rare disease in which mental retardation is associated with abnormal levels of a number of substances in the urine. There is good evidence that it is the result of a failure to produce an enzyme responsible for converting the amino-acid phenylalanine into tyrosine. Phenylalanine is an essential food constituent, but a normal diet contains far more than is necessary for protein synthesis.

Phenylketonuria occurs among the offspring of healthy parents. It is inherited as a Mendelian recessive. However, the heterozygous individuals, although not diseased, can be detected by biochemical tests and hence the doctor can be forewarned that the offspring of two such people could be sufferers from phenylketonuria.

1.4 The component processes of development

Although the cycle of development from zygote to sexually mature adult is often complicated to describe and difficult to analyse, it is helpful to single out some of the key episodes and kinds of process that contribute to it.

1.4.1 Multicellular animals

Gametogenesis (the formation of gametes)

In order to produce a diploid (2n) zygote, fertilization must involve two haploid (n) gametes, that is a *spermatozoon* (a sperm) and an *ovum*, produced by meiotic division from diploid cells.

(a) Spermatogenesis (the production of spermatozoa)

Spermatozoa are almost always small, haploid, motile cells with very little cytoplasm. **[21]** They do, however, contain extranuclear elements which are capable of replication—the centrioles **[22]** and mitochondria. Normally all the products of male meiosis become sperms, a so-called primary spermatocyte gives rise to four spermatozoa. In mammals and many other animals two of each four contain an X-chromosome, two a Y, and upon the chromosome constitution of the sperm depends the sex of the zygote to which it contributes.

(b) Oogenesis (the production of ova)

Ova (eggs) are always large, and in some species vast, cells. **[23]** Their size is partly a reflection of the cytoplasmically stored food resources, or yolk, upon which early embryonic nutrition depends. Meiosis provides one viable egg cell nucleus while the other products die in small cells known as polar bodies. Although many eggs are spherical in shape they are polarized in internal structure (e.g. in the distribution of the yolk and the position of the nucleus) and may show bilateral symmetry. We shall later see that the organization of the egg is of critical importance in early development. Meiosis, in most species, does not proceed to completion until after fertilization. Females are heterogametic in birds, moths and butterflies, and many other animals. In these instances the sex of the zygote depends upon the constitution of the egg nucleus.

Fertilization

In species with small eggs it is usual for only one spermatozoon to enter the egg (monospermy) **[24]**; large eggs may be entered by many sperm (polyspermy) of which only one contributes a nucleus to the zygote. Fertilization stimulates the egg to (a) complete meiosis when necessary, (b) inhibit the entry of further spermatozoa in monospermic eggs, (c) undertake some reorganization of cytoplasmic structure, (d) divide as a prelude to a period of active mitotic cell-division. Fertilization, however, may not be necessary for the development of the egg. Parthenogenesis—development without fertilization—always occurs in species without males (e.g. some lizards), sometimes in other species (e.g. wasps, turkeys, aphids) and can be artificially provoked in yet others (e.g. frogs). **[25]**

21

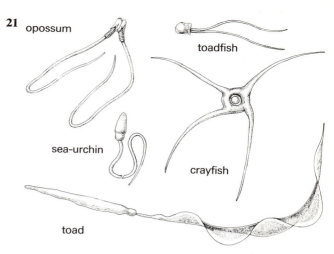

Figure 2 Variations in spermatozoan shape.

22 A centriole is a structure inside the cell which is involved in the separation of the chromosomes during mitosis.

23 The diameters of some of the eggs most studied in developmental work are given below. In these species the egg is approximately spherical:

man	120 μm
mouse	70 μm
frog (*Xenopus laevis*)	1.2 mm
fowl	30 mm

But within a species egg size may be variable. Most somatic cells are much smaller, for example, a mammalian liver cell is about 20 μm in diameter.

24

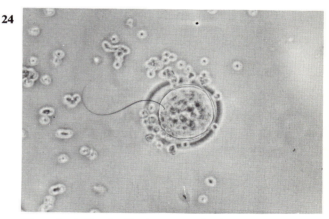

Figure 3 A live fertilized rat egg, showing the 'head' of the spermatozoon in the cytoplasm. (By courtesy of Professor A. Parkes.)

25 ITQ 3

How would you explain the following observations which are known after their discoverer as the Hertwig effect?

Eggs fertilized by spermatozoa that have been treated with ionizing radiation which can damage DNA, develop normally if the dose of radiation is low. As the dose administered to the sperm is raised, a higher proportion of the fertilized eggs develop abnormally and die young. However, above a certain dose this result no longer holds and most eggs complete their early development fairly normally. At a still higher dose no development occurs at all.

Now check your answer against that given on page 28.

Cleavage

Soon after fertilization (or activation in the case of parthenogenesis) the single-celled zygote enters a phase of mitotic activity which leads to the creation of a population of cells. The process is called *cleavage*, the cells produced are called *blastomeres* and the stage of development achieved is known as a *blastula*. The blastula is a hollow ball of cells (from as few as 32 to as many as several millions, according to species) or a distorted version of one. [26] Generally speaking, cleavage sees rather small changes in the relative positions of egg substance; it serves to partition the substance of the egg into cells.

Morphogenetic movement

The cell proliferation characteristic of cleavage continues as development proceeds but is overshadowed by transformations accomplished by massive relative movements of cells. [27] The first of these movements, *gastrulation*, creates an embryo termed a *gastrula* which has an anatomy foreshadowing that of the adult. Layers and groups of cells now come to occupy approximately the positions their descendants will have in the adult organism. The process of modelling the form of the whole, and of its parts or organs, is prolonged. The great movements of gastrulation are followed by less striking minor adjustments. At the end of gastrulation in higher animals it is possible to recognize three major populations of cells, an outer *ectoderm*, an inner *endoderm* and between them the *mesoderm*. [28]

Cell differentiation (or cytodifferentiation)

Cells of early embryos do not closely resemble in appearance, in function, or in chemical composition, the differentiated tissue cells of the adult. The processes by which they come to do so are a striking feature of early development. [29]

Growth

Eggs are smaller than adults. Growth must therefore occur to convert the former into the latter. Growth itself, unless incurring particular restrictions, can lead in time to changes in shape. It is thus a morphogenetic agent.

Genetic control

In some animals (e.g. echinoderms and frogs) the process of cleavage does not seem to involve much synthesis of mRNA and it is fair to assume that the genome is not being used to any marked extent until gastrulation. But as you will see from Unit 3, cytodifferentiation demands the synthesis of some new mRNAs, so the onset of transcription is an important landmark in the history of any cell line.

1.4.2 Multicellular plants

Plants show many variations on the pattern that is fairly standard in animals. The haploid products of meiosis may form, by mitotic activity, large multicellular organisms

26

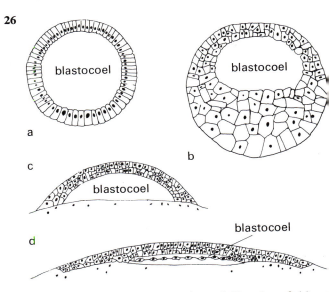

Figure 4 Diagrammatic comparison of blastulae of (a) an echinoderm, (b) a frog, (c) a fish and (d) a bird.

27 Any process in which the form of the organism (or of a part of it) changes progressively and irreversibly is *morphogenetic*.

28

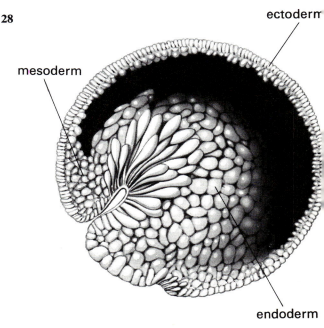

Figure 5 Schematized section through an amphibian gastrula

29 *Differentiation* is a word used in two senses in developmental biology. An originally uniform population of cells is said to differentiate if it gives rise to two or more sub-populations that differ from each other. However, a uniform population of cells which changes in a uniform manner, may also be said to differentiate. This can be shown diagramatically as:

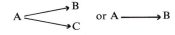

where A, B and C are three different types of cell.

(called gametophytes because they produce gametes). The history of the diploid zygote does not necessarily involve a period in which the whole organism is immature followed by maturity. Plants maintain regions of growth, differentiation and morphogenesis throughout life. Much of a mature tree is dead, but much is undergoing development. It is also a major difference that plant cells are not free to move relative to each other. Change in form in a plant is therefore always linked with growth. Plants offer admirable material for the study of many problems in developmental biology, notably the origin of polarity in cells and the control of differentiation.

1.4.3 Unicellular and other organisms

Unicellular plants and animals may be highly differentiated. Some of the unicellular algae are very large and lend themselves to the kind of experiments that are difficult on other material. Amoebae have rather little by way of constant form or of constant cytoplasmic differentiation, but have been important in nuclear-transfer experimentation (Section 1.8.3). The cellular slime moulds are remarkable in spending much of their life cycle as single cells. Under certain conditions neighbouring cells may aggregate to form a differentiated multicellular body. They are therefore of particular interest in studies on intercellular communication and on differentiation, as you will see from your experiments on slime moulds during the Course.

SAQ for Sections 1.3–1.4.3

SAQ 3 (Objectives 1, 2 and 6)

Which of the following statements are true and which are false?

(a) There is no genetic basis at all to ageing.
(b) In the formation of all blastulae from fertilized eggs there is an increase in size due to an increase in cell number.
(c) Cytodifferentiation occurs in multicellular and some unicellular organisms.
(d) Gastrulae in higher animals contain three cell layers.
(e) In most species spermatozoa are larger than ova.

Now check your answers against those given on page 28.

1.5 Methods of study

What then are the methods by which development can be studied?

1.5.1 The descriptive method

There are limits to the rewards of simply watching intact embryos developing, though in recent years a number of important ideas have come from doing just that or, from what is almost the same, analysing time-lapse photographs of developing systems. For example, the suggestion that gastrulation in sea-urchins is accomplished partly by mesoderm cells throwing out long processes which first anchor to the roof of the blastocoel and then contract to pull the mesoderm in has, as you will see from TV programme 5, been supported by cine-records of the invagination process.[30] Descriptive study may have to overcome natural barriers: for example, in viviparous animals normal developmental processes cannot always be watched. [31]

However, description can be important in answering questions such as: are different rates of cell division in different parts of a developing system responsible for the observed changes in shape which it undergoes? [32] Counts of cells in mitotic division, or estimates of the rate of synthesis of DNA in the nuclei are ways of finding the answer.

Description at the molecular level is also a necessary preliminary to understanding. Red blood cells are full of haemoglobin. When does it first appear in the cells which are their progenitors? When do neurons first begin to transmit impulses? The answers to such questions may come from technically sophisticated investigation, but they are still only describing what happens, in the same sense as does the answer to the question, 'When does the arm first appear?'

1.5.2 The analytical method

The methods of analysis, at the level of both cell and whole organism, are several. We can test the behaviour of some part of the embryo when it is isolated from the rest. The methods of tissue culture allow cells to be cultured outside the body for long periods. [33] Will cells whose *normal* fate is to become neurons do so if they are deprived of contact and association with the rest of the embryo? A great deal has been learned from cell culture about the extent to which developmental processes are carried out in accordance with interactions between cells.

A cell, or a group of cells, can also be transferred or grafted from one part of an embryo to another. Are their developmental fates altered by this? If so, what happens if graft and host are of different species? Such methods were first used at the inter-cellular level of analysis. They can also be used at the cellular level. An egg fertilized by sperm of a foreign species has received a special kind of 'interspecific graft'. If its own nuclear material is removed or killed, we can produce an egg whose cytoplasm comes (mostly) from one species, but whose nucleus comes wholly from another. How would such an egg develop? In recent years a new method (nuclear transfer) has been widely used to study the development of eggs robbed of their own nucleus, but given one from a tissue, or a tumour cell, of an older member of their own species. Even cytoplasm can be grafted. Most of the yolk can be removed from a fowl's egg and replaced by yolk from another egg. The relatively tough outer layer, or cortex, of the amphibian egg has been grafted on to other eggs in place of parts of their own cortex.

Such experiments help us to localize the areas or regions of cell or embryo which are responsible for particular aspects of the behaviour of the whole.

Less easy to analyse are the actual biochemical pathways of importance in development. It is, indeed, easy to show that killing cells stops them developing! What is more difficult is to block a specific synthetic activity and leave the cell alive, but with a restricted or altered capacity to develop. At the present time it is possible to treat cells

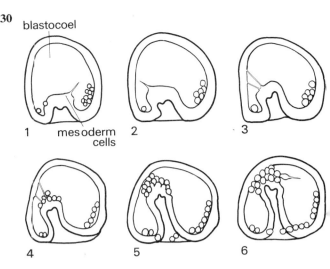

30

blastocoel

1 mesoderm cells 2 3

4 5 6

Figure 6 Schematized diagram showing how the long processes of the mesoderm cells help invagination. 1–6 indicate consecutive stages in this process.

31 Mammalian eggs can be cultured *in vitro* for some time during which they develop apparently normally. Progress in the techniques of egg and embryo culture means that over half the period between conception and birth in rats can be studied *in vitro* (*In vitro* means literally 'in glass' and in this context means outside the intact animal. A cell or molecule inside an intact animal is *in vivo*.)

32 It is important to appreciate that cell division *as such* does not imply growth in total mass or volume of living cells. In many tissues active cell division does indeed accompany growth. But
(a) cells may be removed by death or migration as, or more rapidly than, they are produced;
(b) cells may divide to produce daughter cells without overall growth—this is almost universal in the very early stages of animal development (TV programme 1).

33 Modern tissue culture dates from an experiment performed by Ross Harrison in 1903. He wished to determine whether the axon fibre of nervous tissue was a prolongation of the cytoplasm of a single nerve cell or whether many cells contributed to it. He cultured some embryonic nerve cells outside the body and watched axons sprouting from them. Each axon was the product of outgrowth from a single cell.

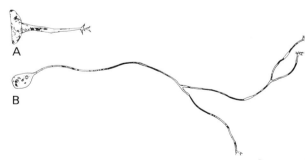

A

B

Figure 7 A: Nerve fibre beginning to grow out from a cell forming part of a small mass. B: A few days later, after the nerve fibre has become branched and greatly elongated.

34 However eggs do have *some* visible structure which is related to their subsequent development. In particular, in nearly all species they show polarity. For example, the position within them of the nucleus is eccentric so that one unique axis runs through the egg in such a way as to pass through its surface

with drugs whose major effect is to inhibit the synthesis of a whole class of macromolecules—DNA, RNA, or protein for example. Even so the drug is unlikely to be absolutely specific. It is also possible to interfere with the metabolism of many of the smaller molecules, which are precursors of the biological macromolecules or are in other ways involved in their synthesis. In Units 2 and 3 you will see something of the molecular changes underlying development.

Now that we have discussed some of the overall processes and techniques of developmental biology, it is convenient to consider some points in a little more detail.

1.6   What changes, and what remains constant, during development?

The early students of development were rightly impressed **35** by the dramatic visible changes in gross size, shape, and structure that accompany development. Later it became clear that at the level of the cell there were equally dramatic changes, and later still the physiology and biochemistry of embryos were shown to undergo fundamental change too. Before we ask if anything remains constant during development, it may be well to consider examples of what changes.

1.6.1   Form and structure

The shape of adults and their gross anatomy may be very complex, but in eggs these features are simple. **[34]** The first important generalization that emerges from study of developmental anatomy is that eggs do *not* transform into adults by the most direct conceivable path. A simple example will make this point. Neither a human egg nor an adult man has a free tail. Yet a human embryo does have one for a short while. **[35]** Similarly, such terrestrial animals as birds do not, as adults, have gill slits, but their embryos for a short period do. These are both examples of developmental conservatism, as human ancestors had tails and avian ancestors were fish. During the development of the ancestors both tails and gills had to be formed and the descendant follows the ancestral path of development more accurately than would be expected if development were as direct as possible. **[36]** However, it is important to realize that embryos, like adults, have to cope with the conditions of their lives, and that developmental conservatism does not prevent the evolution of striking adaptations to the transient needs of embryonic life. Thus the birds and reptiles have evolved a number of structures which serve only to facilitate embryonic nutrition, respiration and excretion within an egg shell that cannot be too porous, or water loss

at the points nearest to and furthest from the nucleus. The distribution of yolk and of pigment in the cytoplasm is a further indicator of real if simple differences between different parts of the egg.

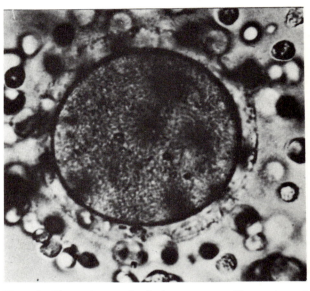

Figure 8   Photomicrograph of a human ovum. (By courtesy of W. J. Hamilton.)

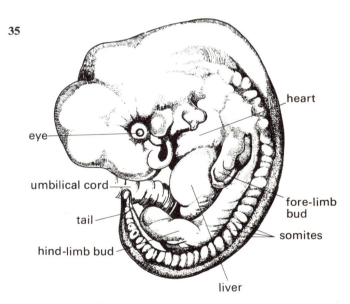

Figure 9   A 30-day human embryo.

36 It follows that a comparison of embryos sometimes suggests an evolutionary relationship that is not at all evident from the comparison of their adults. The most famous such case is that of the echinoderms, the acorn worms, and the chordate animals. Their embryonic anatomy, especially in respect of the coelomic cavities, has similarities which invite a belief in a common ancestry. Modern views support the 'laws' put forward by van Baer (1792–1876), the founder of comparative embryology. They were based only on anatomical findings and run, in paraphrase, as follows:

1   In development general characters appear before special ones.
2   Special characters develop from general ones.
3   During the development of an animal there is a progressive departure from the form of other animals.
4   The early stages of higher animals resemble the early stages, but not the adults, of lower animals.

from the embryo would be lethal. Some of these structures are discarded at hatching. [37]

1.6.2 Function

Before cytodifferentiation and morphogenesis are accomplished, many physiological activities, normal to the adult, are impossible. Blood flow cannot easily precede the formation of vessels and of the heart. Neural control of the heart cannot precede the establishment of its nerve supply. But the development of function does not always take the form of a jump direct to the adult mode. [38] As an example we may take the excretion of nitrogenous by-products of metabolism. These will be formed from the breakdown of, for example, amino acids, purines and pyrimidines. Ammonia is in some ways a good candidate for excretion of nitrogen. However its toxicity at any but the lowest concentrations makes it impracticable unless an exceedingly dilute, and hence copious, urine can discharge it as soon as it is formed. This is possible for animals living in fresh water and tadpoles do, in practice, excrete ammonia. Shell-bound embryos such as those of birds would soon be poisoned by ammonia, but nevertheless they do produce it temporarily, then proceed to urea which is less toxic but very soluble, and finally to uric acid which is sequestered as a harmless, fairly insoluble sludge to be discarded at hatching. [39]

1.6.3 Chemistry

It is impossible to be sure that eggs do not contain minute amounts of substances that are present in great quantities in adult tissues. It is also important to appreciate that an egg or young embryo may contain materials which were synthesized in the mother's body and are passively inherited. Nevertheless, it is quite clear that eggs and embryos differ from adults in overall chemical composition, in synthetic activity and in some metabolic pathways. A few examples will make this plain.

Passive inheritance

An embryo (or foetus or new-born mammal) may contain antibodies which it has not itself synthesized. These offer some protection against pathogens until such time as the young animal can react effectively to antigens by making its own antibodies. Yolk food reserves are a less exotic example of passive inheritance.

Haemoglobin synthesis

Mammal foetuses during most of foetal life have a haemoglobin in their red blood cells which differs in detailed structure from the haemoglobin of adult life. The changeover starts before birth and in rare cases may fail to occur. Foetal haemoglobin is thought to be better adapted to the problems of oxygen transport *in utero* than would be adult haemoglobin. [40]

Metabolic pathways

Before going over to the Krebs cycle, the young embryos of sea-urchins oxidize carbohydrate by different metabolic pathways.

37 The higher mammals are descended from reptiles and have so modified the structures in question that they serve the special needs of life *in utero*. Here there is no problem of water loss and nutrients can be drawn from the maternal blood and excretory products returned to it via the placenta. The shell has disappeared and structures associated with the absorption of yolk (the yolk sac) or with respiration and the storage of nitrogenous waste (the allantois) in reptiles are used to establish close neighbourhood between maternal and foetal blood circulation.

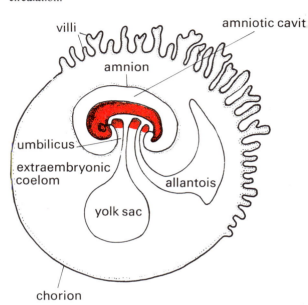

Figure 10 Diagram of embryo (red) of a placental mammal, surrounded by embryonic membranes.

38 Embryos and larvae, no less than adults, show features which are adaptive, that is, features that fit them for survival in their particular conditions of life. Natural selection operates on all age groups. If an embryonic feature (whether anatomical, physiological or biochemical) differs from the adult this may reflect one of two possible situations.

(a) It represents a specific adaptation to juvenile life (e.g. a frog tadpole's tail).
(b) It represents a necessary stage in the development of an adult feature (e.g. a limb lacks nails or claws until it has developed digits).

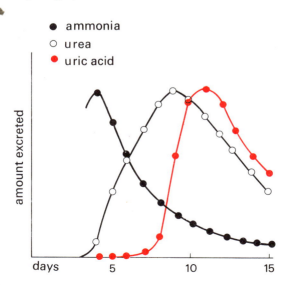

Figure 11 Excretion of ammonia, urea and uric acid during development in a bird's egg.

1.6.4 Exceptions—the constants

With so wide a range of phenomena changing during development it is fair to ask if anything is invariant. The answer is that two important features do not change. First, the egg, though a very unusual cell, is a cell in all essential features and its descendants remain cells in the same sense. Thus the pattern of cellular organization and the major components of cells—nuclei, mitochondria, ribosomes, membrane systems—characterize both the egg and nearly all types of cell derived from it. [41]

The second invariant is chemical and refers to the DNA of the chromosomes. As you know from Unit 17 of S100, the structure of DNA simultaneously permits it to carry specific information in the sequence of bases of which it is largely composed *and* permits this information to be replicated with extraordinary accuracy by the process during which one DNA molecule provides the template for the production of two 'daughter' molecules.

The evidence for the identity, or near identity, of the DNA of the zygote's nucleus and of the nuclei of the adult cells descended from it is of several kinds.

1   The visible behaviour of chromosomes during mitosis is consistent with precise replication. [42]
2   As you will see in Unit 2, nuclei from adult cells can, in one instance, be used to substitute for the nucleus of a fertilized egg.
3   Many of the proteins which are specified by the DNA are the same in different tissues and in an animal of any age. This is true both of proteins tested by their enzymatic activity or by other means. [43]
4   Direct methods of testing the percentage of similar base sequences in two nucleic acid preparations point in the same way.

We shall later (in some detail in Unit 3) wish to consider the molecular biology of nuclear intervention in development. First, however, it is appropriate to see what is known of the processes we are trying to explain in molecular terms. Of these we may take the process of commitment of embryonic cells to their own specific path of cytodifferentiation as a model.

1.7   The process of commitment

The cells of young embryos may differ from each other in size and shape and in their position within the organism. They do not possess the characteristics of adult tissue cells. Sooner or later these must be acquired. Study of the cell-lineage of some organs (e.g. the brain) enables one to say at a very early stage in development that from a particular, localized population of cells, and from no others, will that organ be formed. At the time when such a statement can first be made the cells in question may, individually, be indistinguishable from others in appearance. What is most likely to characterize them is their position in the embryo. How far are appearances deceptive? Are such cells really committed to their normal fate? The answer varies with species and with the stage of development.

In some species blastomeres at the two-cell stage show differences which are associated with the fate of their descendants and cannot be reversed. In others, the blastomeres have entered into quite firm commitments by the

40   The human haemoglobins are more complex, and more interesting, than this suggests. In higher vertebrates the haemoglobin molecule is an assembly of four polypeptide chains with which are associated the oxygen-carrying haems. Characteristically, the four polypeptides in any one molecule are of two kinds. Thus foetal haemoglobin (HbF) has two α-chains and two γ-chains. Normal adult haemoglobin (HbA) has two α-chains and two β-chains. In addition to HbF there are embryonic haemoglobins with other compositions. Quite apart from the different haemoglobins associated with different stages in the life-history, a considerable number of genetic variants are known. (More about haemoglobin structure and function can be learnt from TV programme 3 of S2-1.)

41   Some major departures from normal cellular organization are:
(a)   the red blood cell of the post-natal mammal which has no nucleus;
(b)   multinucleate cells (e.g. certain muscle cells);
(c)   polyploid cells—common in mammalian liver, but found in other tissues also.

Eggs themselves depart from normal cell structure by lacking centrioles. These are normally contributed by the sperm, but may be created *de novo* after parthenogenetic activation of the egg.

42   To this there are a few known exceptions.

43   Other means include chemically separating the proteins and showing their identical structures. (This will be familiar to those of you who have done S2-1.)

8- or 32-cell stage. By this is meant that if they are maintained in any situation compatible with survival and cytodifferentiation they will always follow the same path of differentiation as they would do in the intact embryo. They are said to be *determined*. [44]

In other species, including vertebrate animals, the process of determination is slower, takes place in some regions of the embryo before others, and is progressive. Cells are committed in steps, each one restricting the future possibilities open to them until finally their normal fate is determined. Such progressive determination made the analysis of differentiation easier to study and, in particular, allowed one kind of inter-cellular interaction—*embryonic induction*—to be investigated.

1.7.1 Embryonic induction

Embryonic induction occurs when a cell population that still has some options open to it (e.g. to become epidermal or nervous tissue [45]) has the decision made for it by its contact with another cell population—the so-called *embryonic inducer* or *inductor*. Embryonic inducing systems are responsible for the choice between epidermis and the crystalline lens of the eye [46] and between opaque skin and transparent cornea, as well as being necessary for the formation of the central nervous system as a whole. They are involved in details of the formation of lungs, kidneys, gonads and many other organs.

Embryonic induction shows dramatically that the path of cytodifferentiation adopted by a cell population can depend upon its environment in the shape of contiguous cells. Is the stimulus provided by the inductor specific and does it provide detailed instructions to the induced cells? The answer to the first question is a qualified yes. Although a wide range of non-specific stimuli (damage, heat, centrifugation) can mimic some of the effects of some inductors, it appears that specific chemical agents are involved in the best worked cases. The answer to the second question is no. The inductor is a trigger that sets off a chain of processes in the target cells which must therefore themselves contain the bulk of the information needed to differentiate.

It is characteristic of those inductors that are easiest to test that they do not appear to be specific to a species. [47] Like the vertebrate hormones they will exercise their effects on animals of widely different genotype. The difference between the brain of a man and a mouse may owe little or nothing to the chemistry of the stimuli that commit the cells to form brain tissue in the first place, but may owe most or all to the response of ectoderm cells to the stimulus. [48]

This point is made if we look at another system, the head of amphibians. If at an early stage future belly epidermal ectoderm is transplanted to the head, it will there behave as would normal head ectoderm and form typical local structures. Such behaviour, in which cells adapt to unpredictable change to give a normal end-result, is known as *regulation*. An extreme case is seen in monozygotic twinning where one embryo gives rise to two normally formed ones.

However, as inductive stimuli can work across genetic boundaries, we may ask whether belly ectoderm from one

44 In other words we must distinguish between at least two degrees of commitment. A particular cell, or group of cells, which can be identified in a young embryo may always, in normal development, follow a certain fate; such cells are said to be committed. But if abnormal conditions can change their fate then final determination has not taken place.

45 The induction of the vertebrate central nervous system is an important part of the complex of processes leading to the formation of the embryonic axis. These processes were attributed by Hans Spemann, their discoverer, to the action of an 'organization centre' or 'organizer'.

46 One possible value of inductive relations is that they can ensure a good fit between different components of a compound organ. The lens of the eye must, to work well optically, be in the right position relative to the retina. Its induction by the future retinal cells is therefore apposite.

47

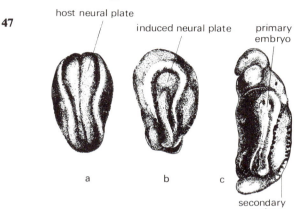

Figure 12 Species non-specificity of inductors.
A piece of gastrula from the so-called *dorsal lip* region was grafted from an amphibian of one species into a young gastrula of another species of amphibian. As can be seen, a secondary embryo is thus induced; a and b are different views of the host and induced neural plate regions. c is a side-view of the embryo at a later stage.

48

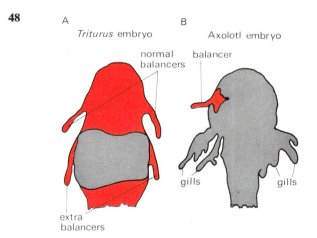

Figure 13 *Triturus taeniatus* normally develops an organ called the *balancer*, another amphibian, the axolotl, does not. In A a piece of axolotl gut was grafted into an embryo of *Triturus* which has as a result developed extra balancers. In B a piece of trunk epidermis of *Triturus* grafted on to the head of an axolotl embryo, has given rise to a balancer (left side of head) while no balancer is formed from the normal axolotl epidermis (right side). The axolotl therefore possesses the required stimulus for balancer formation (A) but its epidermis fails to react to it (B).
Axolotl tissue is shown in grey, *Triturus* in red.

species transferred to the head of another forms host-specific or donor-specific structures. A classical experiment gave the answer: the belly ectoderm of a frog transplanted to the head of a newt provided its host with typical head tissues—but they were typically frog. [49]

This old result is of great theoretical importance as it shows that the responses of developing cells to stimuli from their environment may decide which of a number of possible paths of cytodifferentiation they take, *but* such decisions can only be within the limits of their genetic repertoire. This appears to be generally true of cyto-differentiation. For morphogenesis the most important apparent exceptions concern quantitative features of organs or their parts. The size of a structure is not always typical of the species contributing its cells. [50] We shall return to the topic of embryonic induction in Unit 5.

1.8 Nucleo-cytoplasmic relations

The phenomenon of embryonic induction demonstrates the important influence of interactions between cells during development. But what of the subcellular changes themselves? These involve interactions between the genetic information in the form of DNA in the nucleus and chemical 'signals' emanating from the cytoplasm.

In order to study the interactions of nucleus and cytoplasm, and their relevance to the later history of a cell and its progeny, a number of techniques can be used. [51] Two useful techniques involve *reciprocal hybrids* and *species hybrids*.

1.8.1 Reciprocal hybrids

If a cross between two genetically distinct populations A and B is made both ways, that is, Female A × Male B and Male A × Female B, the two kinds of progeny should be indistinguishable in all respects in which nuclear information only is concerned. If they do differ, the difference may be due to a maternal effect attributable either to the preponderantly maternally derived cytoplasm of the ferti-lized egg or (in the case of mammals, for example) to a maternal environment in post-fertilization life. [52]

In practice, reciprocal hybrids may show some maternal influence, for example, in the rate of cell division during very early development, but usually do not. This does not prove much, however, for two reasons. In the first place the sperm does contain cytoplasm including self-replicating organelles such as mitochondria. In the second place egg cytoplasm may be conveying information which is im-portant but *not* significantly different in the two popula-tions under study.

1.8.2 Species hybrids

Species hybrids, that is hybrids derived from matings between two different species, have certain advantages for work on nucleo–cytoplasmic interactions. It is quite common for small differences between closely related species to be apparent quite early in development. For the most part intra-specific genetic differences are detectable only at late stages of development.

49

50

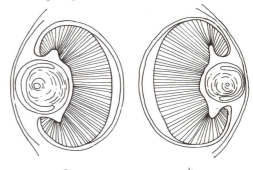

Figure 14 Larva of a newt in which the ectoderm in the mouth region was replaced by ectoderm of a frog embryo. The grafted ectoderm shown in red has developed horny jaws and teeth typical of frog tadpoles.

a   b

Figure 15 a and b show the two eyes of a young tadpole of *Triturus* in which the lens of one eye (a) has been induced out of axolotl ectoderm grafted into the region; the lens is much too large in relation to the eye-cup.

51 Nucleo-cytoplasmic interactions are sometimes referred to as if the nucleus and cytoplasm were separate powers. It is well to remember that the life of either is limited without the other. Mammalian red blood cells, which are naturally enucleated (i.e. without nuclei), can survive for more than 100 days. The unicellular alga *Acetabularia* can survive, and even undertake complex morphogenetic activity, after being robbed of its nucleus. But although enucleate *Acetabularia* is exceptionally long-lived and capable, it too is doomed.

52 Thus C57Bl mice (a highly inbred strain) normally have 6 vertebrae in the lumbar region of the spine while C3H (another strain) have only 5. The reciprocal hybrids follow the mother, but transfer of fertilized ova from the mother to the uterus of another mouse has shown that the effect is due to the uterine environment.

Most attempts to hybridize between species fail, but some succeed [53] and some appear to succeed although the success is illusory. In the latter case the sperm activates the egg but its nucleus does not participate in development, which is therefore parthenogenetic. Sometimes, however, it is possible to remove or destroy the egg's nucleus while successfully fertilizing it with sperm of a foreign species. Any subsequent development will be of an organism the vast majority of whose cytoplasm was originally of one species, but whose nucleus comes from another. Such creatures have been reared through early development in crosses in echinoderms, in amphibians and in silkworms. As the species in question are fairly closely related they differ very little in early embryogeny. Nevertheless, it appears that the nucleo-cytoplasmic hybrid follows its maternal (cytoplasmic) species pattern in such matters as the rate of cell-division during cleavage and gastrulation, but once cytodifferentiation begins it is the paternal (nuclear) species which determines the nature of the embryo.

One celebrated case gave a different result. A nucleo-cytoplasmic hybrid between two newt species was known to die young. By transplanting cells from this hybrid before it died it was found that they could survive well on a normal host of a third species. At metamorphosis (i.e. the change from larval form to adult), weeks after the graft had been made, the grafted cells showed tissue properties characteristic of the species that had provided egg cytoplasm. This result invites further study, but reminds us that cytoplasm may have sources of genetic information of its own.

1.8.3 Nuclear equivalence

Hybridization, referred to above, only allows us to study known or suspected genetic differences between different kinds of organism of the same or different species. But part of our problem is to discover whether or not different cells of the *same* organism have acquired, during development, stable or irreversible differences which might go to explain the differentiated state. In other words, are the nuclei of differentiated cells all equivalent in terms of their genetic information, or not? This can be tested both indirectly and directly.

If the process of differentiation necessarily involved a specific permanent nuclear change, then a differentiated cell could not change its differentiation. Neurons could not become muscle cells, for example. Yet changes in tissue type, called metaplasia, have been reported in many organisms. The most convincing is probably the conversion of differentiated iris margin cells into lens cells in newts. Here, in a process known after its discoverer as Wolffian regeneration, surgical removal of the crystalline lens of the eye is followed by its replacement. The cells that form the new lens are, however, originally quite unlike lens cells—they are, amongst other things, pigmented. Yet though this shows that iris margin cells in the course of their differentiation have not lost nuclear information necessary for lens differentiation, it might be a special case. As you will see from Unit 2, plants provide more telling evidence in that a whole, fertile, adult plant (carrot) can be obtained from the culture of a single isolated tissue cell. The experiments of Briggs and King also led to a more direct approach.

53 Several levels of success can be recognized. It is rare for species hybrids to be fully viable and fertile, though cases are known among wild ducks and flowering plants. More often, if viable, they are sterile (as with mules and hinnies) though they may show what is known as hybrid vigour, whereby the hybrid is healthier than either parent. More often still they die young (sheep-goat hybrids). Many inter-specific crosses in amphibians and echinoderms die during gastrulation—a time at which other evidence shows that transcription of the genome begins in earnest.

As we will see from Unit 3 it is now possible to hybridize somatic cells by persuading them to fuse *in vitro*. In the case of plants this could lead to the formation of complete hybrid organisms since some plants can be grown to maturity starting with a single tissue cell. Somatic cell fusion is, however, often followed by loss of some chromosomes and the cell lines obtained may not be 'complete' hybrids.

Briggs and King were able to remove from a frog's egg the whole of its own nuclear apparatus. Such an enucleated egg normally dies very quickly. But after introducing into it a cell nucleus taken from a somatic cell of another embryo, the resulting egg with a 'foreign' nucleus may survive, and develop. The results of such experiments will be discussed more fully in Unit 2. However, all workers are agreed that it is possible to obtain normal development of a 'nuclear-transplant egg' if the donor nucleus comes from a somatic cell of a very young embryo. It follows that the donor nucleus has been able to provide everything necessary for development that a normal zygote nucleus would.

This conclusion forces us to look again at the organization of the egg cell cytoplasm. It is here that we must seek the origins of the different environments that call forth from the nuclei different parts of their common repertoire.

SAQ for Sections 1.7–1.8.3

SAQ 4 (Objectives 1, 2 and 5)
Which of the following statements are true and which are false?
(a)   Embryonic inductors are not generally species specific.
(b)   Parthenogenesis proves that the sole role of a spermatozoon is to activate the ovum to develop.
(c)   A cell is said to be committed only when, no matter what changes are made in its environment, it develops in isolation as it would in a normal intact embryo.
(d)   Embryonic induction is important in ensuring the correct juxtaposition of different tissues.
(e)   The experiment of Briggs and King shows that cytodifferentiation does not involve a loss of genetic information from the cell nucleus soon after cleavage.

Now check your answers against those given on page 28.

1.9   The internal organization of the egg

The starting point for such processes is of course the egg. Although eggs are anatomically simple they always show *some* structure. Those which are spherical in overall shape are nevertheless polarized in their internal arrangements. The nucleus may lie eccentrically and the point on the surface nearest to it is then known as the *animal pole* (the opposite pole is known as *vegetal*). **[54]** The distribution of cytoplasmic constituents is polarized too. Yolk, pigment, mitochondria, may all be arranged in a defined and constant way in relation to the animal-vegetal axis.

In some species the cytoplasm may in addition possess visible localized concentrations of pigments.

It is clear that such cytoplasmic localizations, if they persist during cleavage, must mean that the nuclei of the young embryo, although themselves equivalent in genetic composition, find themselves in different cytoplasmic environments.

Although the number of different environments may at first be small, once any diversity in cell properties is established it can be used to generate more by cellular interactions. Thus the changes 'snowball'.

54

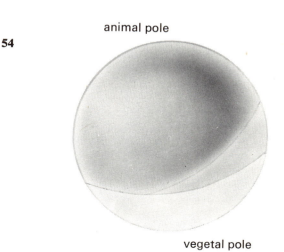

animal pole

vegetal pole

Figure 16   Diagram of a frog's egg.

There is direct experimental evidence that the cytoplasmic localizations of the egg do determine the fates of the cells to which they are allocated. For example, disturbing the distribution of egg materials by mechanical means (usually centrifugation) may cause *corresponding* disturbances in cytodifferentiation. Removal of parts of egg cytoplasm can lead to a later absence of the corresponding embryonic tissue. All this can only mean that the inheritance which an individual receives from its parents is in four forms:

(a) coded, invariant, genetic information in nucleic acid;

(b) cellular organization of a general kind;

(c) specific morphogenetic information in the form of a non-uniform egg cytoplasm which will provide the nuclei of the embryo with different environments.

(d) passive materials of functional, but not necessarily morphogenetic significance (yolk, antibodies, etc.).

It is important to appreciate that although the morphogenetic information stored in the egg cytoplasm is created during its life in the ovary, it may be controlled by the maternal genotype. There are several instances known of the Mendelian inheritance of egg-cytoplasm properties. In these the genotype of the mother, not that of the fertilized egg, determines some feature of development.

Yet although it is certain that we shall discover more and more cases of the effects of maternal genes upon egg structure, there is also reason to suppose that some of the information contained in the egg is not genetically determined in this fashion. For example, in many, but not all, insects the elongated eggs lie in the ovary with their long axes parallel to the long axis of the mother's body. Furthermore, in their subsequent development the polarity of the embryos retains that of the mother—the forward pointing end of the egg in the ovary becoming the anterior end of the embryo. [55] Thus if we consider the egg as providing the starting point for development we find that the egg itself depends on its history in the mother for its subsequent activity. We seem to have recreated the 'chicken or egg' paradox!

55 In animals, other than insects, we still have too little knowledge of the factors determining the polarity of eggs. Other axes of the future embryo are, in some cases, determined at or after fertilization. Thus in frogs' eggs polarity is already apparent at laying, but the future dorso-ventral axis is chosen immediately after fertilization. In the alga *Fucus* polarity in the egg is readily controlled environmentally, by light or by the administration of plant hormones such as indoleacetic acid.

1.10 Summary and conclusions to Unit 1

In this Unit we have tried to present a broad view of developmental biology to show how genotype and environment interact to produce an adult organism from a bud, seed or egg. This has meant the introduction of a large number of systems, ideas and terms which may well be new to you. It is not our intention that from this Unit alone you should become familiar with all these concepts, but you should by now have some understanding of the different component processes that go towards development, and how they are coordinated. Though they are interdependent it is necessary, in order to study and discuss them, to consider these processes in isolation from each other, at least to some degree. In this Unit we considered three central changes that occur in going from egg to adult —*growth*, *cell differentiation* and *morphogenesis*—and how they are influenced by the genotype and the environment. In the rest of this Course we adopt the same strategy but consider each point in much more detail. In Units 2 and 3 we consider the subcellular changes that occur during cell differentiation and how those changes are controlled. In Unit 4 we start to consider the process of commitment and how the path of differentiation of a particular cell or group of cells depends on its interactions with neighbouring cells. This problem is further considered in Unit 5 where we deal with embryonic induction, the nature of the 'signals' between interacting cells, and how cell movement is involved in morphogenesis. Finally in Unit 6, as in the latter part of this Unit (Section 1.9), we are led to examine the structure of the egg in an attempt to understand what sparks off the changes that occur from egg to adult.

Answers to pre-Unit assessment test

1 False   Meiotic division does (19.2.3.)
2 True   (19.2.1.)
3 True   (14.4.)
4 False   (19.6.4. and glossary to Unit 19)
5 False   (19.2.3.)
6 True   (19.2.1.)
7 False   (Appendix 3, Unit 19)
8 False   (17.12.)
9 All three   (19.2.1.)
10 Less than   (19.2.3.)

If you were wrong on more than three questions you are advised to re-read the appropriate Sections of S100, as indicated in the parentheses, before continuing with the text of this Unit.

ITQ answers and comments

ITQ 1 (Objectives 1 and 2)

The earlier the somatic mutation occurs in development the greater its likely effect, since the altered cell will still be destined to go through many rounds of cell division and hence produce a large number of mutated cells. A somatic mutation later in development when cell division is almost over would give rise to an altered cell which would then only produce a few descendent cells.

ITQ 2 (Objective 2)

(a) and (b) would be excluded if thalidomide were administered directly to the foetus, bypassing the placenta, and the foetus subsequently showed specific deformities. Short-term experiments with foetuses cultured *in vitro* are also possible and longer term cultivation of foetal cells could be used to look for changes in cell properties. (c) could be tested directly. (d) could be tackled by administering thalidomide in which particular atoms are radioactive.

However, these suggestions are not the only possible ones. For example, thalidomide itself may be harmless but man could metabolize thalidomide to produce a toxic product while rat left the molecule intact.

This exemplifies the problems associated with attributing 'cause' in biology. To take another example: there are two main schools of thought concerning certain mental illnesses. One school attributes cause to the social and familial conditions of the individual, the other to some biochemical lesion in the individual's metabolism. But can one rather than the other be said to be the 'cause', or is there indeed some other possibility? For instance the social factors might produce the biochemical lesion, *or* the biochemical lesion may make the individual ill at ease with his surroundings, *or* the biochemical lesion, if indeed it exists, may be a secondary effect of the illness and of no relevance to the so-called abnormal behaviour.

Thus speaking of cause in biology should be treated with caution. It is probably all right to attribute the *cause* of say a chemical reaction to one particular enzyme but when dealing with complex systems it is difficult to attribute cause to just one element in that system.

ITQ 3 (Objectives 1 and 2)

The explanation is that:

(a)   low doses cause little damage to the sperm nucleus, so development is normal;

(b)   higher doses damage some or all of the chromosomes in such a way that development is affected;

(c)   still higher doses damage the chromosomes so severely that they cannot participate at all in development which is thus equivalent to parthenogenesis;

(d)   at the highest doses not only are the chromosomes damaged but the whole sperm is killed and unable to fertilize (or activate) the eggs.

SAQ answers and comments

SAQ 1 (Objectives 1 and 2)

Since the resultant zygote will contain 2 sex chromosomes: XY, the child will be male.

SAQ 2 (Objectives 1 and 2)

If red colour is dominant to white, it is possible that each parent plant is heterozygous for the red colour; that is contains one red-producing allele and one white-producing allele of the same gene (r^+/r; where r^+ allele gives the red phenotype, r white). Such plants can each give rise to the same two possible types of gamete containing r^+ or r alleles. Therefore the possible types of zygote are:

$$r^+ + r \rightarrow r^+/r \quad \text{red}$$
$$r^+ + r^+ \rightarrow r^+/r^+ \quad \text{red}$$
$$r + r \rightarrow r/r \quad \text{white}$$

SAQ 3 (Objectives 1, 2 and 6)

(a)   False.   Difficult to prove but, different species have well-defined maximum life spans. Longevity within a species seems 'to run in families'.

(b)   False.   In many species there is no increase in overall size, each cell is smaller than the original zygote.

(c)   True.

(d)   True.   Ectoderm, mesoderm and endoderm.

(e)   False.

SAQ 4 (Objectives 1, 2 and 5)

(a)   True.

(b)   False.   Parthenogenesis allows haploid organisms to develop containing only maternal genetic information. However normal diploid development would also involve paternal genetic information from the spermatozoon. Thus in diploid species the spermatozoon has at least two roles—(i) to activate the ovum and (ii) to provide genetic information.

(c)   False.   Such a cell is said to be *determined*.

(d)   True.

(e)   True.

If you got any of the questions wrong and cannot see why from the answers, you should re-read the relevant Sections of the text.

Acknowledgements

Grateful acknowledgement is made to the following for material used in this Unit:
Fig. 2: Holt, Rinehart & Winston Inc., New York for J. D. Ebert and I. M. Sussex *Interacting Systems in Development,* © 1965, 1970.

28

Units 2 and 3   Cell Differentiation

Contents

Table A

List of scientific terms, concepts and principles used in Units 2 and 3

Introduced in previous Unit	Course Unit No.	Developed in these Units	Page No.
	S100*		
activating enzyme	17		
amino acid	10	asporogenous mutant	48
aminoacyl tRNA	17	basal level	22
antibiotic	10	cellular slime moulds	52
bacteria	17	constitutive	25
cell	14	co-repressor	27
cell division	17	differential gene transcription	35
clone	19	DNA–mRNA hybridization	37
cytoplasm	14	enzyme induction	22
disaccharide	13	enzyme repression	27
DNA	17	histones	31
DNA helix	17	inducer	22
electron microscope	14	inducibility	25
enzyme	15	inducible	22
fatty acid	14	metamorphosis	11
fertilization	19	multicellular organism	9
gene	17	negative control	27
genetic information	17	operator region	25
genetics	17	operon	27
genotype	17	pro-enzyme	23
glucose	13	promoter	25
glycolysis	15	refractility	48
half-life	6	regulator gene	25
hormone	18	repressor	25
hydrogen bond	10	sporulation	46
ion	9	structural gene	25
Krebs cycle	15	temporal control	11
lipid	14	totipotency	12
macromolecule	13	unicellular organism	18
metabolism	15	vegetative cell	46
molecule	8		
mRNA	17		
mutant	19		
neuron	18		
nucleic acid	17		
nucleotide	17		
nucleus	14		
organelle	14		
phenotype	17		
polypeptide	13		
polysaccharide	13		
protein	13		
protein turnover	15		
radioactive	6		
red blood cell	14		
ribosome	17		
RNA	17		
RNA polymerase	17		
skeletal muscle	14		
substrate	15		
thyroid gland	18		
thyroxine	18		
tissue	14		
transcription	17		
translation	17		
tRNA	17		

Table A *(continued)*

List of scientific terms, concepts and principles used in Units 2 and 3

Introduced in previous Unit	Course Unit No.	Developed in these Units	Page No.
ultraviolet light	2		
virus	17		
zygote	19		
S22-**			
cell differentiation	7		
echinoderm	1		
gills	6		
meristem	7		
phloem	2		
sea-urchin	1		
spore	2		
tissue culture	7		
urea	10		
uric acid	10		
xylem	2		
S2-5			
blastula	1		
cytodifferentiation	1		
determination	1		
embryo	1		
embryonic induction	1		
gastrula	1		
parthenogenesis	1		

* The Open University (1971) S100 *Science: A Foundation Course*, The Open University Press.

** The Open University (1972) S22- *Comparative Physiology*, The Open University Press.

Objectives for Units 2 and 3

At the end of these Units you should be able to:

1 Understand the terms, concepts and principles given in Table A.

Demonstrate this understanding by:
(a) distinguishing between the terms, concepts and principles;
(b) distinguishing between true and false statements about them.
(ITQs 3 and 10; SAQs 1, 2, 3, 6 and 8)

2 Describe the principles and general outlines of two experiments that show that differentiated cells are totipotent.
(ITQs 1 and 2; SAQ 1)

3 Give examples to support the notion that cell differentiation involves protein synthesis and that this is regulated.

4 List, or select from a given list, those factors required for cell-free protein synthesis.
(ITQs 3 and 4)

5 Give three reasons why bacteria are simpler to work with than multicellular organisms.

6 Describe the Jacob-Monod hypothesis by drawing and annotating diagrams, or by labelling given diagrams.
(ITQ 8)

7 Evaluate experimental data in the light of, or relating to, the Jacob-Monod hypothesis.
(ITQs 5, 6 and 7; SAQs 3, 4 and 5)

8 Evaluate evidence that assumes a role for control of gene transcription in development.
(ITQ 9; SAQs 6 and 7)

9 Describe two basic lines of evidence that support the idea that control of gene transcription is relevant to development.
(ITQ 9)

10 Evaluate evidence that assumes a role for control of translation in development.
(ITQ 11; SAQ 7)

11 Evaluate evidence about the relative roles of nucleus and cytoplasm during cell differentiation.

12 Describe briefly, using a simple diagram, the life-cycle of a sporulating bacterium.

13 Evaluate evidence concerning hypotheses to explain the temporal control of bacterial sporulation.
(ITQ 11; SAQ 8)

14 Describe briefly, in the form of a simple diagram, the life-cycle of a cellular slime mould.

15 Give the advantages and disadvantages of using bacteria as model systems for cell differentiation in higher organisms.

Study guide to Units 2 and 3

These Units consider cell differentiation. They assume a knowledge on your part of S100, particularly of Unit 17. If you are not familiar with that Unit you are advised to re-read it, particularly Sections 17.2–17.10 inclusive and Appendix 1 (Black). You should check your knowledge of Unit 17 (before, after or instead of re-reading it) by attempting the pre-Unit assessment test given on p. 8. You should then check your answers to this test with those on p. 54, and do the appropriate re-reading indicated by those answers before starting these two Units. Throughout these Units you will find in-text questions (ITQs) and self-assessment questions (SAQs). You will probably benefit most by doing these (and checking your answers) as you come to them. It may prove impossible to understand the text following some of the ITQs without having first attempted these questions or at least having looked at the answers. You need remember only those names of chemicals, etc. indicated in Table A; however, you may find that remembering some specific examples other than those in Table A helps you remember the principles they illustrate.

A word or two about the TV programmes: TV programme 2 relates particularly to Sections 2.4.1, 2.4.2 and the home experiment. You will most benefit from the programme if you have already studied up to the end of Section 2.4.2. TV programme 3 relates specifically to Section 3.2 so it will help you to read to the end of that Section before seeing the programme.

Those of you who have studied the Biochemistry Course, S2-1*, will find some of Unit 2 already familiar to you. Sections 2.4, 2.4.2, 2.5 and 2.5.1 cover similar ground to Unit 6 of S2-1. This is because their subject forms a bridge between biochemistry and developmental biology. We therefore advise you to study this topic again by reading Sections 2.4, 2.4.2, 2.5 and 2.5.1, even if you know it well. Similarly, those of you going on to do S2-1 after S2-5 should not omit reading Unit 6 of S2-1. Those of you who have not done S2-1 and are perhaps not going to, need not worry, as Sections 2.4, 2.4.2, 2.5 and 2.5.1 need no prior knowledge of biochemistry other than that obtained from S100.

* The Open University (1972) S2-1 *Biochemistry*, The Open University Press.

Pre-Unit assessment test

You should attempt this test, before proceeding with these Units.

Questions 1–10 Indicate which of the following statements are true and which are false.

1   Enzymes are composed mainly of protein.

2   Ribosomes are the major sites in the cell at which ATP is synthesized.

3   DNA is a double helical molecule in which each strand of the helix is composed of a polypeptide chain.

4   The messenger RNA is formed by copying the sequence of nucleotide bases from a strand of DNA.

5   Transcription is the process by which messenger RNA is decoded on the ribosomes to form polypeptide chains.

6   All the organisms in a clone are genetically different.

7   Thyroxine is a hormone which is produced by the pituitary gland.

8   The DNA of cells of higher organisms is located mainly in the nuclei of those cells.

9   The phenotype of an organism is unaffected by the environment.

10   On the genetic code, for each amino acid there is at least one specific codon.

Question 11 From the list of components below, choose those which are necessary inside the cell for the synthesis of protein:
(a) ribosomes; (b) ATP; (c) glucose; (d) Ca^{2+}; (e) tRNA; (f) amino acids.

Question 12 Name two substances other than those from question 11 which are necessary for protein synthesis to occur within cells.

Question 13 Name three factors which can cause mutations to occur.

Questions 14–18 Indicate which of the following statements are true and which are false.

14   A mutant is a genetically altered form of an organism.

15   A multicellular organism is composed of a large number of identical cells.

16   A zygote is a cell which results from the fusion of two gametes.

17   The process of biological evolution cannot operate without genetic variation.

18   In higher organisms the genes are located in the chromosomes.

Now check your answers against those given on p. 54.

Unit 2

2.1 Introduction

As you know from S100, all higher organisms are multicellular, that is, they are complexes consisting of a large number of individual cells. Yet all these organisms, whether animal or plant, derive from a single cell, such as a fertilized egg (Unit 1). As the adult organism is larger than the single cell from which it comes, development must involve growth as well as cell division (TV 1). Growth and cell division alone would be expected to produce a roughly spherical ball of identical cells. This is not generally sufficient to produce a viable adult organism, as not all the cells in an adult multicellular organism are the same, as microscopic examination soon reveals (Fig. 1).

Figure 1 Different types of cell: (a) skeletal muscle cells; (b) a neuron; (c) a portion of an intestinal epithelial cell (S100, Unit 14).

(a)

(b)

(c)

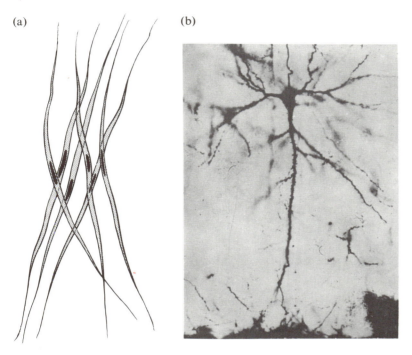
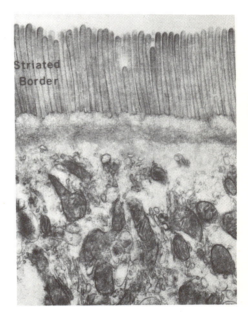

Therefore, at some stage during the development of a multicellular organism from a single cell via successive cell divisions, the resulting cells must change in character to yield finally all the different types of cell found in the adult organism. This process, whereby different cell types are generated, we have termed *cell differentiation* or *cytodifferentiation* and it is the topic under consideration in these two Units. Of course, in the adult organism, the different cell types are not merely arranged at random; they are clustered in specific regions to form tissues and organs. Indeed, we usually name a cell according to the tissue or organ in which it occurs. This spatial arrangement of the cells is probably in part a consequence of cell differentiation and in part responsible for cell differentiation. A cell which is of a particular type may during development tend to move to, or stay in, a particular site in the embryo, *or* because a cell is in a particular site in the embryo it may change in character (differentiate) accordingly, as for example in the phenomenon of embryonic induction mentioned in Unit 1. Embryonic induction implies that cells pass and receive 'signals' to and from neighbouring cells. These 'signals' influence the path of cytodifferentiation taken by any particular cell. The nature of the 'signals' themselves presumably depends on the stage of differentiation of the cells from which they emanate. Thus here we can see that the path of cytodifferentiation taken by any particular cell depends on how it influences and is influenced by its neighbouring cells. As these processes are interdependent, it is somewhat difficult to consider them separately; however, for the sake of clarity we shall do so. In these two Units we shall restrict our discussion to the subcellular changes that occur to cells during differentiation and how these changes are controlled. In Units 4 and 5 we shall consider how the position of a cell in a developing embryo affects its differentiation and *vice versa*, and what the nature of the 'signals' between cells might be. Since we are not concerned here with the spatial relationship between cells, we can conveniently represent the process of cell differentiation as a 'lineage chart' as in Figure 2 (over page).

cell differentiation

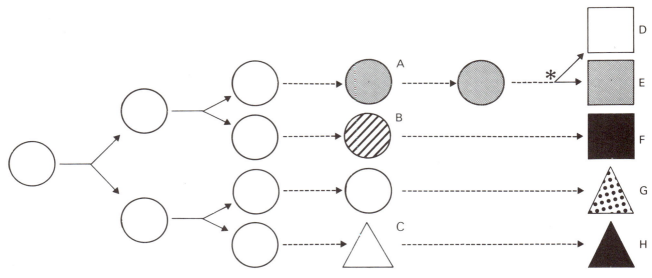

Figure 2 indicates the origins of each cell type (D–H) during cell differentiation via cell division from the original single cell (zygote).

There are two points to note:

1 Some cell types that arise during development are not present in the adult organism (A, B and C).

2 At branch-points marked with an asterisk, each daughter-cell of that division can become either of two types of cell (e.g. D or E).

Each cell is presented with a 'choice'. What in essence we wish to consider is: what governs this choice and how is it made? By 'choice' we are not suggesting that a cell has a mind, but are using this word as a way of expressing the situation facing a cell at a branch-point. Why does it go one way or the other? Having gone one way, how does that cell change? We ask these questions now, but as you will discover by studying these Units they have not yet been fully answered. So what we are going to do is to show how, by many years of study, these questions have been refined to a point where they are sufficiently precise for one at least to see *how* to attempt to answer them! To begin, we take two features of development as axiomatic:

1 The zygote contains information for making all the different types of cell characteristic of that organism. A zygote is therefore said to be *totipotent*.

2 The process of cell differentiation is controlled.

The information in the zygote must be sufficient to generate all the cell types present in the adult organism, and intermediate cell types that arise during development. This information is by definition 'genetic information', for example, the information for making a frog in a fertilized frog's egg is passed on from one generation of frogs to the next. As you know, this genetic information is called the *genotype* of the organism. Within a given species of organism the genotype will be broadly the same; a frog is always a frog, etc. However, minor differences in genotype can occur, leading for example to differences in eye colour or, somewhat more dramatically, differences between male and female. During development there is an interaction between the genotype of the organism and the environment in which it is expressed. This interaction yields the final adult organism (*phenotype*). So different environments can affect development differently (Unit 1) though only within fairly narrow limits: a fertilized frog's egg cannot be made to develop into a rabbit, or even for that matter into another species of frog. Drastic alterations in the normal environment tend to be lethal or at least to arrest normal development severely so that an adult fertile organism is not attained. You will, of course, already be familiar with this concept of genotype–environment interaction from Unit 1, but consider what we mean by environment when we are looking at differentiation of any particular cell or group of cells in an embryo. The environment for those cells is both the environment exterior to the embryo as a whole and the effect produced by the cells surrounding those under consideration. We emphasize again that these cell–cell interactions are very important in development (Unit 1); they are considered in detail in Units 4 and 5.

Figure 2 Cell lineage. Each circle, square, etc. represents a type of cell. Non-differentiated cells (i.e. like the original zygote) are open circles. A junction is taken to represent cell division. Dotted lines indicate an unspecified number of intervening cell divisions. D, E, F, G and H represent the five cell types present in the adult hypothetical organism. After the first few divisions, many cells of each intermediate type will be present.

totipotent

Control of cell differentiation obviously occurs, as does control of growth. The production of the correct cell types in the correct numbers, even leaving aside their occurrence in correct positions, is hardly likely to occur by a totally random process. Control of timing of cell differentiation, so-called *temporal control,* also occurs, as the production of different cell types (Fig. 2) happens in a specific order and at specific times during development. This temporal control of the use of the genetic information present in the zygote is dramatically seen in organisms that show metamorphosis, that is, organisms where development involves a free-living larval stage. For example, a butterfly egg must contain information for making both its larva (the caterpillar) *and then* a butterfly from this caterpillar. All the changes in cell type that this demands must occur in the correct time-sequence. So, tying together our two axioms: cell differentiation depends on a controlled use of the genetic information in the zygote. This use depends on the information itself and its interaction with the environment. The interaction occurs in both directions: 'signals' from the environment affect the information and may lead to some changes in the cells, while changed cells imply a change in the environment for their neighbouring cells.

temporal control

In these two Units we shall consider the control of the subcellular changes that occur during cell differentiation, only briefly mentioning the 'signals' (Unit 5). To do this we first consider the nature of the genetic information and what happens to it during development (Section 2.2) and then the nature of the differences between cell types (Section 2.3). Having in this way defined more exactly the connection between the genetic information and its cellular manifestations, we go on to consider bacteria, which gives us some clues about the possible control mechanisms involved in the changes that occur to differentiating cells (Section 2.4). We then examine the relevance of these findings to higher organisms (Section 2.5). Unit 3 begins with a consideration of how relevant the control mechanisms discussed in Unit 2 are to cell differentiation (Sections 3.1–3.3). These Units end by considering some interesting experimental systems that may assist in answering our current questions (Sections 3.4–3.5).

2.2 The nature and distribution of genetic information

> STUDY COMMENT In this Section we consider the fate of the genetic information during cell differentiation. You are not expected to remember details of the experiments of Steward and Gurdon, but should be able to recall the rationale behind them (Objective 2).

The differentiated cells in an adult organism are specialized. That is, each type of cell is constructed in a way that enables it to fulfil a particular function; thus a neuron has long processes to conduct electrical impulses, a skeletal muscle cell has actinomyosin fibrils to allow contraction to occur, and so on (Fig. 1). During normal 'wear and tear' in an adult organism, cells need to be replaced and this replacement is achieved by cell division, thus generating two cells from one; Where a specialized cell divides, it gives two similarly specialized daughter-cells; a liver cell gives two daughter liver cells. That is, the phenotype of a specialized differentiated cell is stable. In contrast, although a zygote is also specialized for its job, the cells that are the products of its division can, via successive cell divisions, give rise to all the different types of differentiated cell, characteristic of the organism in question. To do this the zygote must contain the genetic information to enable it to generate all the cell types; we have taken as axiomatic that the zygote is *totipotent*. But what of a specialized differentiated cell that divides to give only similar differentiated cells, which themselves divide to give more of the same type of cell, and so on; does that cell, with its limited potential, contain the same wide complement of genetic information as the zygote? Consider an analogy.

In building an aircraft one of the essential steps is to construct a master plan. This must contain the overall design of the aircraft, detailed information about the individual parts and the programme for building and assembling those parts to complete the job. This master plan is like the genetic information in the zygote. Now, in building the aircraft from this master plan, not all the parts can be built and assembled simultaneously. Different parts must be built by different groups of craftsmen—one group to build the engine, another to make the tyres, still another to make the seats, and so on. If the master plan contains all the details, there are two ways of organizing the building of the whole aircraft. One way is to distribute to a particular group of craftsmen a copy of just that portion of the master plan that they need to do their part of the construction. Alternatively, each group can be given a copy of the complete master plan; in this instance the master plan must also contain additional instructions to each particular group about what part of the overall plan *they must act on*. So, going back to specialized differentiated cells: does each type of cell contain a copy of a limited portion of the initial zygote's information or, does each type of cell contain all the genetic information, which includes 'instructions' as to which portion that type of cell must obey? These alternatives are shown schematically in Figure 3.

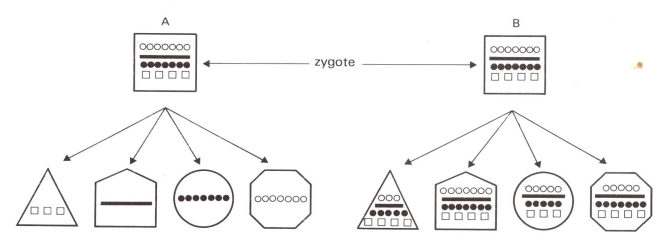

These questions are of fundamental importance to the whole conceptual framework within which we view the process of cell differentiation. They have in part been answered, over the last 20 years, by experiments that have radically influenced the direction of research in the field of cell differentiation, and are likely to do so for many years to come. So it is to these crucial experiments that we now turn.

Figure 3 The distribution of copies of genetic information. Types of cell are represented by circles, squares, etc. Different genetic information is represented by different types of line. Scheme A: Only parts of the zygotic information are received in each type of cell.
Scheme B: Each type of cell receives a copy of the complete information in the zygote.

2.2.1 Totipotency

If scheme B in Figure 3 is a correct representation of the distribution of genetic information during cell differentiation, then one could make some interesting predictions about the potential behaviour of differentiated cells. If we could take a single cell from a differentiated tissue and place it in a chemical medium in which the cell could undergo cell division, it might be possible that the cells generated by many subsequent divisions would differentiate into other additional cell types and, eventually, yield a whole organism similar to that from which the original cell was obtained. This would depend on two conditions being fulfilled:

1 Scheme B is correct.

2 We have the ability to provide the correct 'environment' for the single differentiated cell to simulate normal development. This we would not expect to be easy. After all, it is not particularly easy to get even zygotes to develop outside their normal environment (e.g. only limited development has yet been achieved of mammalian embryos 'in a test-tube'). Of course, if scheme A is correct then no amount of tampering with the experimental environment could result in the single differentiated cell yielding a whole organism.

In practice, one can in fact isolate single cells or fragments of tissue from a wide variety of animal or plant organs and get the cells to divide in chemical media. This division is often quite efficient and large amounts of cells can be grown and maintained in this way. The composition of the media required varies from tissue to tissue and is usually determined in the laboratory by a process of trial and error. Cells have been kept or 'cultured' in this way for many years and the technique is commonly known as 'tissue culture'. As this name suggests, cells isolated from a particular tissue tend to divide to give cells like those in the tissue from which the original cells derived. For example, individual muscle cells grown in culture in a glass dish can eventually fuse together to give a muscle-like tissue which can even contract! Though changes do occur to cells in tissue culture, one does not generally get organized differentiation into different cell types. However, as the aim of the investigators is often to maintain cultures with all the

tissue culture

characteristics of the tissue from which the cells originated, the media have been developed mainly to prevent cell differentiation. Thus, all these studies do is to reiterate the stability of the phenotype of differentiated cells. They do not allow us to distinguish between schemes A and B, as the inability to get a whole organism from one differentiated cell may merely reflect that condition 2 has not been satisfied.

However the story does not end there.

Plant cell culture

In 1948 Caplin and Steward were studying the growth of small pieces of carrot phloem in culture media. Growth was slow as cell division was very limited. In trying out various chemical media they found that cell division could be rapidly accelerated by adding *coconut milk*! This is not as crazy or as lucky as it may seem. Coconut milk is a form of liquid endosperm, the endosperm being a tissue that is present in, and important to the development of young plant embryos. Further additions to the medium, such as certain plant hormones known to be important in normal plant development, further improved the growth rate of the carrot tissue. Nevertheless, the tissue did not differentiate to give a whole plant. Some time later Steward and his associates turned their attention to carrot cultures based on single isolated carrot cells. Once again using their coconut milk medium, good cell division could be achieved; a suspension of about 5 cells per cm^3 could give rise to one of about 100 000 cells per cm^3 in 2 to 3 weeks. However sometimes after a cell division the two daughter-cells did not separate and when they divided, their daughters did not separate either. In this manner nodules of cells arose. From these nodules then arose structures with roots, shoots and leaves! These structures, which Steward called 'embryoids', in turn give rise to plantlets and eventually to mature normal carrot plants (see Fig. 4). This outstanding experiment could be repeated on a whole variety of plants and, furthermore, in some species of plants, single cells taken from any one of a variety of tissues (stem, root, etc.) could all give rise to whole plants. Thus differentiated cells in the plant, like the original zygote, appeared to be totipotent! Scheme B is proven—or is it?

Figure 4 The growth of a carrot plant from a single cell.

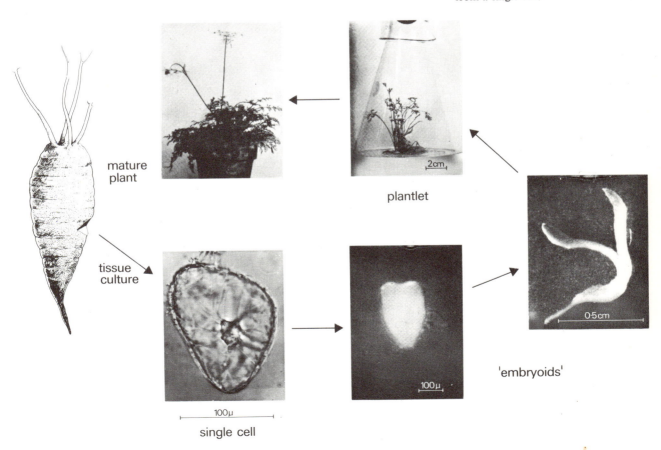

mature plant

plantlet

tissue culture

single cell

'embryoids'

ITQ 1 From your knowledge of growth and differentiation in plants (S22-, Unit 7), does this experiment prove that all types of differentiated carrot cells are totipotent? Can you think of any exceptions?

You should check your answer against that on p. 54 before reading on.

There are further points worth noting from Steward's classic experiments:

1   Whether an isolated plant cell divides to give a culture of only like cells or eventually to give a mature plant depends on the culture conditions.

2   Different types of differentiated cell from the same plant need different culture conditions to allow them to yield whole plants.

3   Only when isolated as single cells, not as a mass, will differentiated plant cells grow, divide and give rise to mature plants.

4   Cells growing as single cells do not resemble cells from the tissue from which they were isolated.

The importance of these observations should be more apparent later in this Course. In brief, they suggest that, though totipotent, a differentiated plant cell has a stable phenotype partly because of its association with its neighbouring cells (3). Removal of these associations can under appropriate conditions (2) allow a differentiated cell to express its full informational potential like a zygote. This expression may involve a prior 'de-differentiation' (4), that is, a stage where the cell 'loses' many of its specialized characteristics. Somewhat similar losses of special features are sometimes observed in tissue cultures of animal cells; but as yet nothing as clear cut as Steward's results has been achieved with animal cells.

Does this mean that differentiation of animal cells is basically unlike that in plants and operates via scheme A (Fig. 3)? After all, we do know that plants (unlike animals) contain regions of growth and differentiation throughout their life. Or does this inability to form whole animals from a single differentiated cell merely reflect that the experimenters have yet to find the animal equivalent of coconut milk? A somewhat different line of experiments bears on these questions.

Nuclear transplantation

By using careful techniques it is possible to remove the nucleus from an amoeba cell. Without its nucleus the animal cannot divide and soon dies. This and similar studies (Unit 1, side note 51), show that the cell nucleus is vital to development. This is of course consistent with the fact that the DNA of the cell is mainly located in the cell nucleus and, as you know from Unit 17 of S100, it is this DNA that we define as the genetic information. So considering Figure 3 in terms of DNA content we would argue that according to scheme A each type of differentiated cell nucleus contains a copy of only part of the DNA present in the zygote nucleus. On the other hand, according to scheme B all the cell nuclei would contain a complete copy of all the DNA. As you saw in Unit 1 (Section 1.6.4) there is some evidence that the DNA content of the cells does not vary during differentiation, though this evidence cannot rule out the possibility that small, yet highly important, differences do occur. Now consider the following experiments:

As you saw from Unit 1 (side note 23 and TV 1), the eggs of amphibia are large cells. Using micro-dissection techniques, one can remove the nucleus from a fertilized frog's egg and watch the subsequent development. In fact, as with amoebae, the enucleated egg will not develop and soon dies. It is possible, however, to replace the nucleus with one from another cell. In the 1950s experiments of this kind were performed by Briggs and King. As a source of nuclei they used cells from frog embryos at various stages of development; that is, the donor cells (those from which the nuclei to be transplanted were obtained) were at various stages of differentiation. They then observed the subsequent progress of these 'nuclear-transplant eggs' (i.e. zygote cytoplasm plus embryo cell nucleus).

ITQ 2 Assume for the moment that the 'host' zygote cytoplasm would, in principle, act as a 'coconut milk' for the 'donor' nucleus and allow it to express its full genetic information. If scheme A (Fig. 3) is correct, what would you expect the relationship to be between the degree of development of the nuclear-transplant egg and the stage of development of the embryo from which the donor cell nucleus was isolated?

You should check your answer against that on page 54, before reading on.

As you learnt from Unit 1, if one uses cells from very young embryos, as donors of nuclei, then the nuclear-transplant egg can develop normally. This, however, may merely reflect the lack of differentiation of the cells of the young donor embryo. In fact, when working on the frog *Rana pipiens,* Briggs and King found that donor nuclei from blastulae or gastrulae would support some development in the nuclear-transplant egg. However, if the donor nuclei were from later embryos, little or no development occurred. This therefore appeared to support scheme A (see answer to ITQ 2).

However there are some important objections to these experiments:

1 In removing the nucleus from the zygote the remaining 'host' cytoplasm and cell membrane may have been damaged.

2 The nuclei of later embryos may be more subject to damage than those from earlier embryos, as the cells of later embryos are smaller and therefore harder to handle.

The technique of nuclear transplantation is certainly a tricky one. However, using another species of frog *Xenopus laevis,* Gurdon has managed to achieve complete development of nuclear-transplant eggs even using donor nuclei from embryos later than gastrulae. His technique has the advantage of inactivating the zygote nucleus with ultraviolet light rather than surgically removing it from the egg. In some cases, Gurdon has achieved complete development of nuclear-transplant eggs through to adult frogs, using donor nuclei from the intestinal cells of tadpoles. It appears that the genetic information in the nuclei of differentiated cells in animals is the same as in the zygote nucleus. So in animals, as in plants, in terms of genetic information, differentiated cells are totipotent. (There are of course, some exceptions as certain differentiated cells (such as mammalian red blood cells) like plant xylem cells, differentiate to the extent of losing their nuclei.) The failure to achieve development of single animal cells to give whole organisms, as was shown to be possible for plants by Steward, may be due to technical problems plus the possibility that animal cells do not readily de-differentiate. The demonstration that differentiated cells in both animals and plants are totipotent poses another question: how is the stability of the phenotype of a differentiated cell normally maintained? Some clues to the answer can be derived from the findings of Steward and Gurdon.

At this point you should attempt to summarize the main findings of these two experimenters as described above, and then suggest what they tell us about the stability of the state of differentiated cells and cell nuclei. Do not spend more than about 15 minutes on this. Then compare your summary with that below. If they differ substantially you should re-read Sections 2.2 and 2.2.1.

Summary of Steward's and Gurdon's findings

1 Differentiated cells in an organism are totipotent. That is, they contain the same genetic information as the zygote.

2 Dissociation of plant tissue into single cells 'triggers' de-differentiation; then a subsequent differentiation to give whole plants is possible.

3 Though stable in a differentiated cell, the nucleus of a differentiated animal cell when placed in a zygote cytoplasm can sometimes express its full genetic information to give a whole adult organism.

Points 2 and 3 suggest that the phenotype of a differentiated cell is maintained through, in part at least, two influences:

(i) As dissociation of plant tissue into single cells seems to be needed to trigger de-differentiation, the contacts between neighbouring cells normally present in pieces of tissue must be important in maintaining the differentiated states of the cells.

(ii) As nuclei from differentiated cells are equivalent to the zygote nucleus in terms of genetic information, the zygote cytoplasm must differ from that of the differentiated cells. Presumably the cytoplasm of a differentiated cell helps maintain the nucleus in a state where only a limited amount of its genetic information is used and hence the differentiated cell phenotype is stable.

This implies that in the process of cell differentiation changes occur in both the nucleus and the cytoplasm of the cell because of 'signals' within the cell and from outside the cell. Inside the cell these 'signals' are interpreted so that they alter the use of genetic information. Different cell types must use different portions of the information. In other words, use of the genetic information in the zygote is differentially switched on or off during the process of cell differentiation. We will consider the nature of these switching mechanisms in the remainder of these two Units.

SAQ for Sections 2.1–2.2.1

You should now attempt the following question:

SAQ 1 (Objectives 1 and 2) Which of the following statements are true and which are false?

(a)   The experiments of Gurdon show that differentiated animal cells, unlike plant cells, contain less genetic information than the zygote.

(b)   The genetic information in a cell is in the form of DNA.

(c)   A totipotent cell is one that has only a limited content of specialized genetic information.

(d)   Cell–cell interactions are important in maintaining the stability of the differentiated state.

(e)   The activity of the cell nucleus to some degree depends upon the cytoplasm in which it is located.

(f)   De-differentiation involves a destruction of genetic information.

2.3   Differences between cells

In attempting to find out how the use of particular genetic information may be switched on or off, it is first necessary to establish what this information is actually used for. The short answer is of course 'for building an organism' or in terms of cell differentiation 'for making different types of cell'. But what in detail does 'type of cell' mean? In what ways does one type of cell differ from another? The answers to these questions provide clues about how genetic information is used.

Figure 1 demonstrated that cells can vary in relatively gross ways—in size, shape and the type or number of organelles (mitochondria, ribosomes, etc.) that they contain. But these differences must ultimately depend on differences in the chemical composition of cells.

Consider any cell in any multicellular organism. Then make a list of the broad classes of chemical compounds that it could be expected to consist of. Do not spend more than 5 minutes doing this. When you have finished compare your list with that given below.

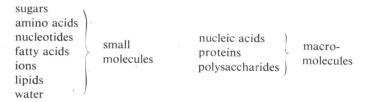

This list can be subdivided into two: small molecules and macromolecules. The small molecules either come directly from the environment surrounding the cell or from the various metabolic inter-conversions that take place within the cell (S100, Unit 15). These inter-conversions are all dependent for their rapid rates on specific biological catalysts: enzymes. Macromolecules within any cell are generally synthesized within *that* cell and their synthesis too depends on enzymes. Thus, it is no exaggeration to say that the overall chemical composition of the cell substantially depends on the enzymes that it contains—the type of enzymes and their amounts. Enzymes are proteins. Other proteins, which are not enzymes, help form various cellular structures such as membranes. Therefore the character of any type of cell is largely determined by the spectrum of proteins (structural proteins and enzymes) it comprises. Thus we would expect cell types that differ in a gross way (Fig. 1) also to differ in the proteins they contain. In fact, all cell types within any one organism contain many proteins in common, as all cell types share similar metabolic pathways and similarly structured organelles. However, they do show some differences in certain proteins and it is these proteins by which they differ that help lead to the gross differences. These differences in the protein content of different cell types must arise during cell differentiation. Whether these changes are the cause of cell differentiation or just the effect is not really easy to say nor is it particularly meaningful. Indeed we have already mentioned the problem of speaking of 'cause' in highly complex inter-acting systems such as those that occur in biology (Unit 1). We shall just examine what changes do occur during cell differentiation and the mechanisms underlying these changes. Later on, by reference to specific systems, we will tackle the somewhat more meaningful issue of whether some specific changes in protein composition that occur during cell differentiation are necessary for subsequent changes to occur. We are fairly sure, as pointed out above, that the gross changes associated with cell differentiation can be explained in terms of changes in protein composition.

One can detect certain specific changes in the protein composition of specific cells associated with particular times in cell differentiation. For example, as you saw in Unit 1 (Section 1.6.2), tadpoles, being aquatic creatures, excrete waste nitrogen in the form of ammonia. Frogs excrete their waste nitrogen as urea. Thus, on metamorphosis from tadpole to frog, the animal must develop the mechanisms for producing urea.

These mechanisms involve a number of enzymes which are found almost exclusively in the liver cells and these enzymes rise in amount (i.e. the amount of enzyme protein per cell) during metamorphosis (see Fig. 5).

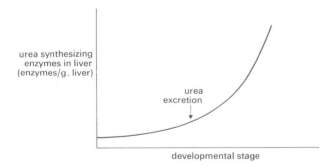

Figure 5

So we can perhaps regard cell differentiation as involving the gradual changing of the protein content of various cells during the process of development: in an early embryo where all the cells are the same they contain the same proteins, but as development proceeds different cells acquire different spectra of proteins. How does this arise and what mechanisms control it?

As you know from Unit 15 of S100, even in an adult organism proteins are 'turned over'. That is, proteins are broken down and replaced by newly syn-thesized proteins. So the protein content of any cell, both in terms of types and amounts of proteins, depends on the rates of both synthesis and degradation of proteins. *Cell differentiation could therefore involve differential alterations in the synthesis and degradation of specific proteins during development.* We can now link together the switching on or off of the use of genetic information with cell differentiation. Consider in summary the following points:

synthesis and degradation

1   Different cell types contain different proteins.

2   Different cell types contain the same DNA (genetic information).

3   DNA is transcribed and translated, as described in Unit 17 of S100, to give proteins. One unit of genetic information, a gene, yields one protein or poly-peptide chain. One gene is one particular sequence of nucleotide bases along a stretch of DNA helix.

Putting these points together we can propose that the differences in the protein composition between cells that occur during cell differentiation arise in one of three possible ways:

I In different cells different genes are transcribed so that they yield different spectra of proteins and hence different types of cells. Protein degradation is insignificantly slow or the same for all types of protein.

II Different cells vary in both the genes that are or are not transcribed and the rate at which the protein products of these genes are degraded.

III In different cells all the genes are transcribed so that they yield the same spectrum of proteins *but* different proteins are degraded at different rates in different cells.

These three hypotheses are compared in schematic form in Figure 6.

What we now wish to examine is whether any of the above hypotheses really explains the changes seen during cell differentiation In other words we wish to answer two questions: (a) what are the relative importances of specific degradation and synthesis of proteins? (b) what are the mechanisms by which these specific processes operate and are controlled?

When trying to answer such complex questions it is generally desirable to use an experimental system in which the phenomena under examination can be controlled by the experimenter. Cell differentiation is a complex process occurring as part of an even more complex process, development. For this reason, most studies of what factors control changes in the cellular levels of proteins have been carried out not on developing systems but on adult organisms where changes can be brought about in just a few proteins under carefully controlled conditions. The rationale behind the experiments is that the mechanisms underlying these changes will probably be similar to those underlying the changes that occur during cell differentiation. For example, changes in the levels of several enzymes in the liver of a rat can be effected by changing the animal's diet from one containing a small amount of glucose to one containing a large amount of glucose. Similarly, injection of certain hormones into an animal can lead to dramatic changes in the levels of specific proteins in specific tissues. The diet changes or injections are of course controlled by the experimenter, and the effects occur rapidly and do not depend on long periods of growth and development. Even with these advantages it is often very difficult even to establish the answer to question (a) as protein synthesis and degradation both occur in the tissues of higher organisms. It is experimentally tricky to distinguish between these two ways of changing protein levels: for example, if an increase is seen in enzyme E, is this due to an increased rate of synthesis of E *or* a decreased rate of degradation? It would be convenient to use a system where only synthesis or degradation was occurring at any one time so that the mechanism of that alone could be examined. Unlike higher organisms, bacteria, which are unicellular organisms, do not exhibit a significant rate of protein degradation when growing rapidly. This can be readily demonstrated by growing bacteria in the presence of radioactive amino acids. During this growth they synthesize proteins and these will carry the radioactive label. The radioactive amino acids are then removed from the medium. The subsequent loss of radioactivity from these proteins (due to their breaking down and being re-synthesized using non-radioactive amino acids (S100, Unit 15)), is found to be very slow. Under some circumstances bacteria do however exhibit changes in the level of certain proteins and so these changes must be due to increased protein synthesis. For this and, as you will see, for other reasons most of what we know about the control of protein synthesis has come from studies on bacteria.

2.4 The control of protein synthesis in bacteria

Before we discuss the results obtained over the last 20 or so years on the control of protein synthesis in bacteria, it is useful to consider what features determine bacteria as the right sort of organisms for such studies. (Some of these arguments will have been presented on the Biology Radio Programme, 'Choosing the right organism'.*)

* For details, see the *Broadcast Notes*.

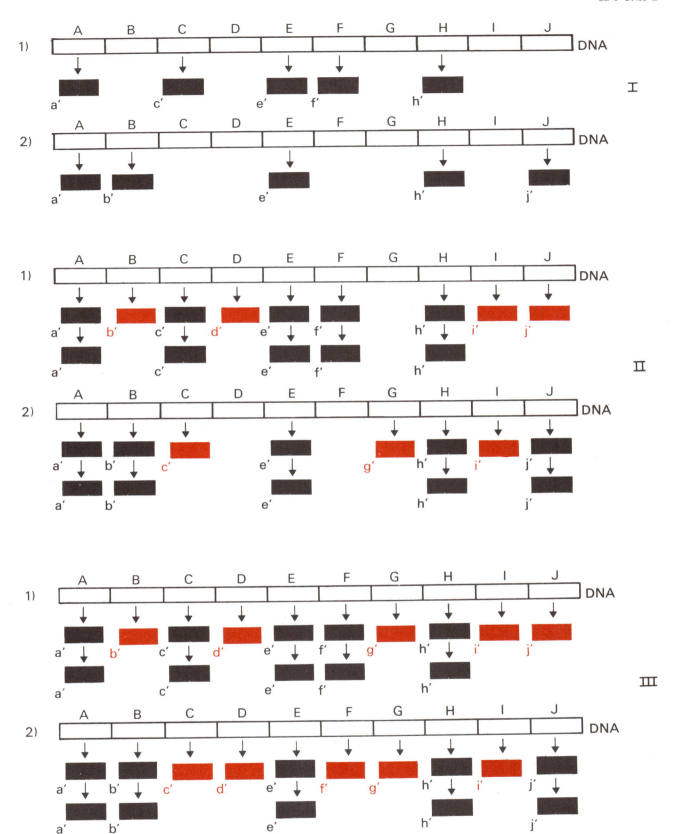

Figure 6 Hypotheses I, II, and III for generating two different types of cell, 1 and 2.

The capital letters (A–J) refer to different genes, the small letters (a′–j′) to the corresponding proteins. (None of the intermediary mechanism for synthesizing proteins based on information in DNA is shown.) Proteins written in black are considered stable, those in red as readily destroyed in the cells in question The final protein spectra of cell types 1 and 2 are the same, irrespective of hypotheses I, II or III. Type 1 contains proteins a′, c′, e′, f′, h′; type 2 a′, b′, e′, h′, j′.

2.4.1 Bacteria–biochemists', geneticists' and molecular biologists' delight!

When choosing an organism as a system in which to study a particular biological problem at the cellular level, we believe the following features are desirable:

1 Obviously the organism must exhibit the phenomenon being investigated.

2 Large amounts of the cell tissue under study should be available.

3 If genetic analysis is required then large numbers of individual organisms are needed, and large 'litter sizes' are useful.

4 The organism should be easy and cheap to handle.

Compare bacteria and, say, a mammal such as a mouse in relation to these four points and our current problem.

Point 1 Changes in the levels of particular proteins can be shown in both organisms. Bacteria have the specific advantage in this case that little significant protein degradation occurs within them.

Point 2 Bacteria reproduce by simple cell division. In many species of bacteria one organism (i.e. one cell) gives rise to two within 30 minutes; two to four within 60 minutes, and so on. Therefore kilogram quantities of bacteria can be grown within days. If you were working on, say, mouse liver cells, you would require hundreds of mice for similar quantities of liver cells. The mating of two mice yields about half a dozen offspring in about 20 days. In this time, one bacterium would theoretically give rise to about 10^{289} cells or approximately 10^{274} kilograms of cells, a mass many times greater than that of the universe.

Point 3 The arguments are similar to those in point 2. Furthermore all the bacteria derived from one original cell placed in a culture medium are virtually genetically identical individuals. A group of genetically identical individuals, termed a 'clone' of organisms, is useful for genetic and biochemical studies. Two sibling mice are only virtually genetically identical if they derive from the splitting of a single zygote (monozygotic or identical twins). Thus thousands of millions of genetically similar bacteria can be obtained, while by inbreeding mice (e.g. mating brother and sister, etc.) only a few relatively genetically dissimilar individuals are obtained. Thus genetic differences have to be taken into account when using several individual higher organisms. On the other hand, it is often convenient when attempting to analyse normal mechanisms to compare normal individuals with genetically abnormal ones (mutants). As the desired abnormalities may be very rare it is convenient to be able to obtain and examine many individual organisms. With bacteria, techniques have been developed to enable one to 'choose' a required mutant from amongst many thousands of millions of individuals—imagine breeding that many mice! The ease with which such mutants can be obtained in bacteria should be apparent from this week's television programme (TV 2).

Point 4 To produce enough mice for just one experiment may take several weeks during which the animals need feeding, warmth, cages, places to store them and people to clean them. Enough bacteria for an experiment can be grown overnight. The bacterial 'cage', food and storage, as you know from your home experiment, is a test-tube of nutrient-containing agar jelly. If you need further convincing of the advantages of working on bacteria, watch TV programme 5 of the Biochemistry Course (S2-1),* which we urge you to watch anyhow! Of course, before bacteria can be regarded as useful organisms for studying the control of protein synthesis with a view to understanding its control in cell differentiation in higher organisms, one other criterion must be met: the basic mechanisms of protein synthesis in bacteria and higher organisms should be the same. Fortunately, they are, though they may differ in detail.

SAQ for Sections 2.3–2.4.1

SAQ 2 (Objective 1) Which of the following statements are true and which are false?

(a) Bacteria reproduce by simple cell division.

(b) A gene comprises enough genetic information for making one polypeptide chain.

---

* Details are given in the *Introduction and Guide to the Course*.

(c)   The number of nucleotide bases in a gene determines the number of different types of polypeptide chain made from that gene.

(d)   Protein degradation is more rapid in bacteria than in higher organisms.

(e)   Protein in the cells of higher organisms survives as long as the cells do.

(f)   The DNA of a mutant bacterium is identical in every respect with that of the original parent bacteria.

.4.2   Enzyme induction in bacteria

STUDY COMMENT   This Section contains material familiar to those of you who have studied Unit 6 of the Biochemistry Course (S2-1). We nevertheless suggest that you study this Section carefully as the arguments are crucial to the rest of these Units. Some additional experiments bearing on the Jacob-Monod hypothesis are also discussed in TV programme 2.

You need not remember the names or chemical formulae given unless directed to do so, as in Objective 1 (Table A).

If bacteria are 'Royal Organisms' to many experimental biologists then one particular species of bacterium, a dweller in human intestines, *Escherichia coli*, is 'King' among them. This simple organism having found an ecological niche in the gut of man has more recently found another in the test-tubes of biochemists and geneticists alike. The reasons are again simple: it is easy and cheap to keep and grows quickly. More is probably known about the biology of *E. coli* than any other single organism. *E. coli* grows and divides rapidly (about every half-hour) when provided with a simple medium comprising NH_4^+ ions (as a source of nitrogen), K^+, Mg^{2+}, Na^+, SO_4^{2-} (as a source of sulphur), PO_4^{3-} (as a source of phosphorus), small amounts of some other metal ions, and a source of carbon. The source of carbon can be any one of about 20 different organic compounds. The best source, that is, the one on which *E. coli* grows most rapidly and therefore the most commonly employed in laboratories, is glucose. On this sugar *E. coli* divides about once every 30 minutes when grown at 35 °C. The glucose is converted by *E. coli* via the glycolytic pathway, Krebs cycle, and all the interconnected metabolic pathways to provide the carbon necessary for synthesizing all the organic compounds that comprise *E. coli* (S100, Unit 15). All these metabolic conversions require specific enzymes which *E. coli* also synthesizes, and this again requires carbon.

Consider, however, what happens when we provide *E. coli* not with glucose but with another source of carbon, say lactose. Lactose is a disaccharide consisting of one molecule of glucose linked to one molecule of galactose via a β-galactoside link (S100, Unit 13). To metabolize lactose *E. coli* must first break this β-galactoside link and it can do this by means of a specific enzyme, β-galactosidase. The glucose and galactose formed can then be further metabolized via the glycolytic pathway.

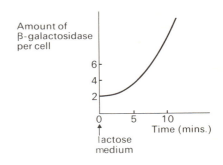

Figure 7   Induction of β-galactosidase.

lactose $\xrightarrow{\ \beta\text{-galactosidase}\ }$ galactose   +   glucose

Glycolysis

When *E. coli* is growing on a medium containing glucose as the sole source of carbon, the cells contain little β-galactosidase. If some of these *E. coli* are removed from the glucose medium and placed into a medium in which the sole source of carbon is lactose, some dramatic changes take place in the level of β-galactosidase in those cells (see Fig. 7).

21

The glucose grown cells were transferred to lactose at time 0. The original level of β-galactosidase (at time 0) is called the 'basal level'.

This increase in the level of β-galactosidase per cell is known as *induction of β-galactosidase*. The substance causing it, lactose, is called the *inducer* and the *E. coli* cells which are able to manifest the increase are said to be *inducible*. When the increase occurs they are said to be *induced*. Many other examples of enzyme induction in bacteria are known wherein specific enzymes are induced by their specific substrates. The general name for this phenomenon is *enzyme induction*. (Take care not to confuse this with 'embryonic induction' referred to in Units 1 and 5. Enzyme induction always refers to increases in enzymes, embryonic induction to changes in cell types. Unfortunately there has been a tendency over the years to abbreviate both terms to 'induction'.)

inducer

enzyme induction

The evolutionary advantages of enzyme induction are obvious; a cell makes a high level of an enzyme only when the substrate *on which that enzyme acts is present*. This represents a considerable economy to the organism. Thus, if lactose is removed from induced cells (and replaced by, say, glucose) or broken down by those cells, no further increase in the level of β-galactosidase occurs. The β-galactosidase present does not break down much as protein degradation in bacteria is slight. However, the level of β-galactosidase *per cell* will decrease at each cell division, that is, if no further β-galactosidase is being synthesized, then as each cell divides to give two the level per cell drops by 50 per cent, at the next division by a further 50 per cent, and so on. If lactose is again added to such cells further induction can occur, and so on. There is in fact a 'ceiling' to the induction, though with certain analogues of lactose, induction can result in up to a thousandfold increase over the basal level of β-galactosidase; the induced level can represent up to about 6 per cent of the total cell protein. As protein degradation is slight in bacteria, this huge increase in the level of β-galactosidase or similarly induced enzymes, cannot arise from a reduction in the normal rate of protein degradation. It must arise from an increase in the rate of synthesis of that specific protein. To understand how this increase occurs and is controlled, we must now review what is known about the basic mechanism of protein synthesis, which as we have said, is essentially the same in all organisms.

The mechanism of protein synthesis

In Unit 17 of S100 you learnt about the process by which the information encoded in the sequence of nucleotide bases in DNA is decoded to determine the sequence of amino acids in polypeptide chains which then fold up to give proteins. As a way of recapitulating these facts we suggest that you try the following ITQ.

> **ITQ 3** In Figure 8 (opposite) we show diagramatically the synthesis of three specific proteins based on information in three adjacent genes in DNA. We have arbitrarily subdivided the process into three steps: one step is transcription, one protein-folding, the other one translation. Which step corresponds to each number in the diagram, (1, 2, 3)? In the list below we give the names of the chemicals or structures needed for these three processes. You should subdivide this list into three parts, depending on the steps in which the substances listed are needed. Some of the substances listed are shown in the diagram as A, B, C, D. Identify them.
>
> *Chemicals or structures needed for protein synthesis*
>
> Messenger RNA (mRNA), Transfer RNA (tRNA), amino acids, activating enzymes, ribosomes, nucleotide bases, aminoacyl tRNAs.
>
> *When you have finished (do not spend more than 15 minutes), you should check your answers with those on p. 54, before reading on.*

If you examine carefully the complete diagram on p. 55 (answer to ITQ 3) you will notice that there are many different substances required in the synthesis of any protein. Alteration of the level of any of these could in principle result in a change in the rate of protein synthesis. For example, if an organism suffered a reduction in its cellular level of ATP, its rate of protein synthesis would be expected to fall, as would be expected from reduced levels of any of the 20 amino acids, and so on. However enzyme induction is characterized by the fact that it is specific, which means that in response to any particular inducer only one protein (or sometimes a few related proteins) increases in level. This must

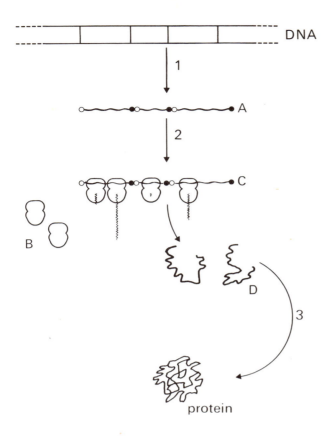

protein   *Figure 8*

mean that some component involved in protein synthesis which is specific to one or a few particular proteins must be affected. In any particular case there are only three such components.

> **ITQ 4** Look at Figure 32 on p. 55. Name the three components involved that are specific to specific proteins.
>
> *Now check your answer with that on p. 55.*

So enzyme induction could involve one or more of three separate steps in protein synthesis. For β-galactosidase, for example, it could involve:

(a)   increased transcription for the gene for β-galactosidase to give more mRNA for β-galactosidase (called β-galactosidase mRNA);

(b)   increased rate of translation of existing β-galactosidase mRNA to give more of the polypeptides that comprise β-galactosidase;

(c)   more efficient folding up of these polypeptides to give more active β-galactosidase enzyme: (In the 1930s this was thought to involve a mechanism like that shown in Figure 9).

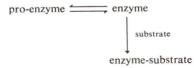

We are now going to analyse some data that bear on these three alternative hypotheses. But before doing so, it is worth examining some additional experimental observations and some of the consequences of each hypothesis.

Figure 9 A combination of enzyme with its substrate (inducer) was supposed to shift the equilibrium between polypeptide chains (pro-enzyme) and enzyme, so as to produce more active enzyme.

The start of the increase in β-galactosidase on adding inducer to the cells is very rapid (see Fig. 7). Likewise, this increase ceases almost immediately when inducer is removed. Therefore the switch-on/switch-off is very rapid. If (a) is correct and the mechanism depends on switching on or switching off the transcription of the gene for β-galactosidase, then the β-galactosidase mRNA must be unstable, so that when further mRNA ceases to be synthesized because transcription of the gene for β-galactosidase is switched off the existing mRNA is soon all gone. Hence further synthesis of β-galactosidase rapidly ceases. If (b) is correct then mRNA could be stable or unstable (as long as it is continually synthesized), as it is the rate of translation of the mRNA not its amount that is, according to this hypothesis, supposed to be controlled. If (c) is correct the stability of the mRNA is irrelevant but an uninduced cell (i.e. one to which no inducer has been added) must contain a 'pool' of β-galactosidase polypeptides (i.e. a store of pro-enzymes).

Over the last 30 years the lactose induction system has occupied many scientists. This work has yielded many important results and has allowed a decision to be made for this system between alternatives (a), (b), (c) above. But it is not just for this one system in this one organism that these results are so important, as the studies led to a whole new hypothesis about how protein synthesis might be controlled. This hypothesis formulated in 1961 by two French biologists, François Jacob and Jacques Monod, revolutionized our ways of thinking about protein synthesis and indeed cell differentiation. Their work, and that of others, involved both biochemical and genetic techniques, a happy marriage within the auspices of the 'Church of Molecular Biology'. We cannot, of course, discuss 30 years' work but we shall briefly indicate in this text and in TV programme 2 the main lines of the research and how it allowed the unravelling of the lactose system, how this led to their hypothesis and how the outstanding ideas of this hypothesis have now been justified.

Jacob-Monod hypothesis

The unravelling of the lactose system

The basic observations are as follows:

1 On addition of lactose, or some non-metabolized chemical analogues of lactose, to *E. coli* cells, an increase is observed in the cellular level of β-galactosidase and two other enzymes related to lactose metabolism (called galactoside permease and transacetylase). These increases start within a few minutes. (Fig. 7).

2 Some analogues of lactose, though not substrates of β-galactosidase, were found to be inducers.

3 Puromycin, a substance that inhibits the joining together of amino acids to form polypeptides, when added to *E. coli* along with lactose prevents any induction of β-galactosidase.

4 If radioactive amino acids are added to *E. coli* along with lactose, and the β-galactosidase subsequently induced is isolated from the cells and purified, it is found to be radioactive.

> **ITQ 5** From these first four observations, what can you conclude about alternatives (a), (b) and (c)?
>
> *Now check your answers with those on p. 55.*

5 It is possible, by genetic techniques, to construct a map of the relative positions on the DNA of the three genes containing the information for the three enzymes of the lactose system. These genes are found to map consecutively on the DNA and are transcribed to give one long mRNA molecule that is translated in, as it were, three regions to give the three enzymes (Fig. 10).

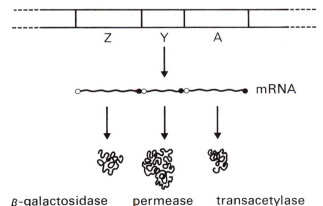

Figure 10

6 The half-life of β-galactosidase mRNA was found to be very short, of the order of 1–2 minutes.

7 Many mutants of *E. coli* were examined. Some were found that made a high level of the three enzymes even in the absence of inducer. These mutants were called *constitutive,* as they made the enzymes as part of their constitution. By genetic techniques it was shown that these mutations all occurred in a separate gene, located near to the Z, Y and A genes. This gene, called *i* for inducibility, was said to be a *regulator gene,* as it is involved in the regulation of the amounts of the enzymes synthesized and not in their protein structure. Indeed, examination of the β-galactosidase from constitutive mutants shows it to be identical in structure with that from normal inducible cells. By contrast, Z, Y and A are called *structural genes.*

constitutive

regulator gene

structural genes

    ITQ 6 What can you now say about alternatives (a) and (b)?

    Check your answer with that on p. 56.

Jacob and Monod considered evidence of the type given above along with evidence gathered from other, mainly genetic experiments, such as the so-called PAJAMO experiment discussed in TV programme 2. From their considerations they concluded that:

(i) The *i* gene contains information for making a protein called *repressor*. This repressor exists in small amounts inside the cell.

repressor

(ii) Repressor, in the absence of inducer, specifically inhibited the synthesis of the three enzymes.

(iii) Inducer, when present in the cell, binds specifically to repressor.

(iv) Repressor, with inducer bound, is incapable of inhibiting synthesis of the three enzymes.

They further guessed that alternative (a) was correct and that the repressor acted by inhibiting transcription. Inhibition of transcription would mean a rapid switch-off of synthesis of the enzymes as the mRNA was unstable (see point 6, above). They and their co-workers invoked from further genetic studies the existence of two additional important regions on the DNA—the *promoter* (P) and the *operator region* (O). P was the region on the DNA before a set of genes where RNA polymerase, the enzyme responsible for copying the DNA sequence into mRNA, attached. O was a region where, in the absence of inducer, repressor could attach. As O was between P and the structural genes (Z, Y and A) when a repressor was attached to O, RNA polymerase could not get to, and hence could not transcribe, the structural genes in mRNA. These points of their hypothesis are summarized in Figure 11 (over page). Study it carefully and satisfy yourself that it is compatible with all the above data on the lactose system.

promoter
operator region

In summary: repressor (coded for by the *i* gene) in the absence of inducer binds to O and as RNA polymerase attached to P cannot reach Z, Y and A, no mRNA corresponding to these genes is made (Fig. 11A). When present, inducer removes repressor from O (Fig. 11B). RNA polymerase can now reach Z, Y and A and transcribe these genes to give mRNA, which is then translated to give the three enzymes (Fig. 11C).

    ITQ 7 Though compatible with this hypothesis, does the data given in 1-7 prove it?
    Also, is (a) or (b) correct?

    See the answer on p. 56 now.

More recent data has made almost certain that the guesses made by Jacob and Monod in 1961 are correct:

1 Following the addition of inducer to *E. coli* an increased amount of β-galactosidase mRNA can be detected. Inducer seems to cause transcription of mRNA to occur.

2 Repressor has recently been isolated. That is, a protein has been isolated that has been coded for by the *i* gene and hence made by inducible cells but not by constitutive ones.

3 This isolated repressor has been shown to bind specifically *in vitro* to DNA isolated from the operator region, but not to DNA from other regions. This binding is prevented by inducer.

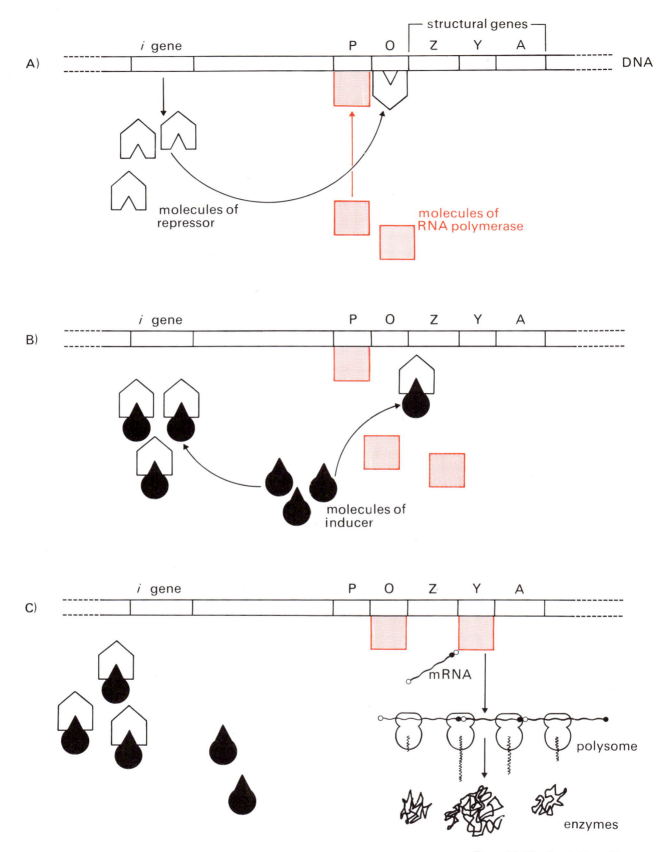

A)

i gene   P   O   structural genes   Z   Y   A   DNA

molecules of
repressor

molecules of
RNA polymerase

B)

i gene   P   O   Z   Y   A

molecules of
inducer

C)

i gene   P   O   Z   Y   A

mRNA

polysome

enzymes

4   From *E. coli* cells it is possible to derive a cell-free system comprising DNA, RNA polymerase, ribosomes, tRNA, activating enzymes and all the small molecule substrates required for protein synthesis (answer to ITQ 3). Such a system will synthesize β-galactosidase. This synthesis is prevented by adding isolated repressor.

In general, Jacob and Monod termed a system where the regulation of the synthesis of several enzymes is coordinated and the structural genes for these

Figure 11 The Jacob-Monod hypothesis. P is the promoter, the region o the DNA where the enzyme responsib for copying DNA sequence into RNA (transcription), RNA polymerase, attaches. O is the operator region.

enzymes are consecutive on the DNA, an *operon*. Thus for each operon there is a specific regulator gene (e.g. *i* gene for the lactose operon) that contains the information for a repressor. This repressor is specific for the operator region of the operon and the inducer. Thus, control on their scheme is *negative*, that is, the rate of transcription is set by the degree of inhibition by the repressor. This hypothesis can also be readily adapted to explain a related phenomenon in bacteria, *enzyme repression*. In this phenomenon, the synthesis of certain groups of enzymes related through their metabolic roles is decreased on addition of certain small molecules (called *co-repressors*) to the cells. (That is, it is the reverse of induction where the small molecule (inducer) allows enzyme synthesis.) Once again, the control is negative and is considered to be exercised by specific repressors. For any one repressible operon there exists a regulator gene which contains information for a specific repressor. This can bind to an operator region and hence inhibit further transcription, *only in the presence of* the small molecule co-repressor. (Conversely, in induction the small molecule, inducer, prevents repressor binding to O.) Existing mRNA, (i.e. made before addition of co-repressor) of short half-life, soon decays.

Enzyme repression is summarized in Figure 12.

Figure 12 RNA polymerase is shown in red. In (A), transcription is occurring; two molecules of RNA polymerase are shown here, one just having reached the operator region (O), the other having by now travelled halfway along the structural genes. In (B), transcription has ceased owing to repressor–co-repressor having bound to the operator region. So in repression, repressor can only bind to the operator region (thus inhibiting further transcription) *when the small molecule co-repressor is present and bound to the repressor*. This contrasts with induction where the small molecule (the inducer) when bound to the repressor *prevents it binding to the operator region*.

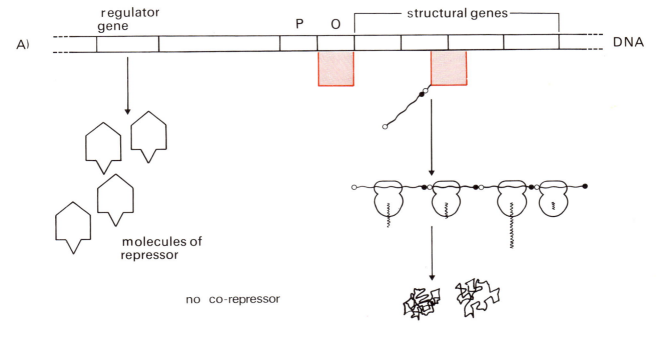

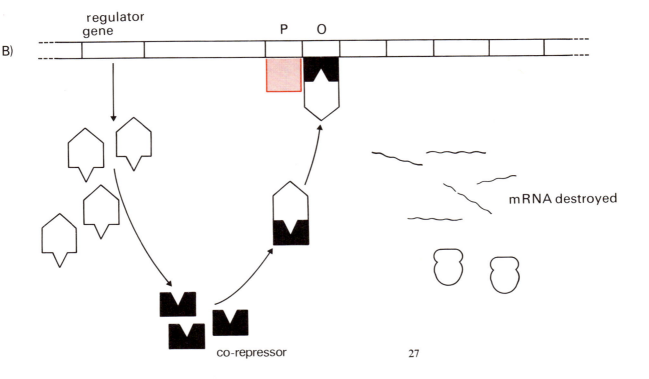

We should also note that not all enzymes are subject to induction or repression; some are made constitutively. Generally, these are enzymes which are in central metabolic pathways, such as glycolysis, and as such are always necessary at more or less constant levels. Their actual levels are probably geared to the overall growth rate of the cells.

Though only really proven for the lactose system in *E. coli*, the Jacob–Monod hypothesis has been applied to many superficially similar phenomena of induction and repression in many organisms. However, as you will see even where induction or repression do occur, the assumption that all such phenomena occur in the same way as the induction of β-galactosidase is somewhat precipitous. So much so that this tendency has been punned, admittedly horribly, 'Monodenous'. The main achievement of Jacob and Monod is not merely in explaining the control of synthesis of β-galactosidase in *E. coli*. More importantly they have provided a framework within which many studies relevant to the control of protein synthesis in general, and indeed to cell differentiation, can be initiated. It is this stimulus to further research that was aptly recognized in their share in the Nobel Prize for Physiology and Medicine in 1965.

We now suggest that you try to summarize the main points about induction (2.4.2), particularly the central features of the Jacob–Monod hypothesis. Then check your summary with that given below. If they differ substantially you should read this Section again.

Summary of Section 2.4.2

1   Induction of certain enzymes occurs when substances that are substrates of those enzymes or analogues of substrates are present in the cells.

2   Induction begins within minutes.

3   Induction involves synthesis of new polypeptide chains, not just folding of pre-formed ones.

4   mRNA is unstable in bacteria.

5   Control of polypeptide synthesis could in principle operate by controlling the rates of transcription or translation.

6   The Jacob–Monod hypothesis visualizes that:

(a)   control is via regulation of transcription.

(b)   the structural genes for related enzymes which are induced or repressed together are adjacent on the DNA, that is, they constitute an operon.

(c)   each operon has a specific operator region, adjacent to the structural genes.

(d)   each operon is regulated by a specific regulator gene.

(e)   each regulator gene contains information for a specific repressor.

(f)   the repressor binds to its specific operator region, thus preventing the access of RNA polymerase to the structural genes.

(g)   the binding of the repressor to the operator region is prevented by the inducer (or, in repression, helped by co-repressor).

7   Some enzymes are always made constitutively, since their amounts are not influenced by the level of substrates (or co-repressors) in the cells.

SAQs for Section 2.4.2

SAQ 3 (Objectives 1 and 7)   Which of the following statements are true and which are false?
(a)   Enzyme repression involves the binding of a repressor to a regulator gene.
(b)   Lactose is an inducer of β-galactosidase because it provides the *E. coli* cells with glucose.
(c)   An operon comprises several adjacent structural genes and an operator region.
(d)   The joining together of amino-acids to form polypeptide chains is catalysed by RNA polymerase.
(e)   Each one of the 20 different types of amino acid has a tRNA specific to it alone.
(f)   A promoter is a region on the ribosome where tRNA binds.
(g)   A regulator gene controls the structure of several enzymes.

SAQ 4 (Objective 7) On adding the amino acid serine to some *E. coli* the cells show an increase in the enzyme, serine deaminase, an enzyme that degrades serine. If this increase occurs by mechanisms similar to those controlling the induction of β-galactosidase, which of the materials shown below, if added to the cells along with serine, would you expect to (a) enhance the increase in serine deaminase, (b) reduce it, (c) have no effect on it?
(i) The amino acid, tryptophan. (ii) Na²⁺ ions. (iii) puromycin, an inhibitor of the joining together of amino acids to form polypeptides. (iv) naladixic acid, an inhibitor of DNA synthesis. (v) rifampicin, an inhibitor of RNA polymerase. (vi) deoxyglucose, an inhibitor of ATP synthesis.

2.5   Induction and repression in higher organisms

Having discussed the control of protein synthesis in bacteria we can now consider the subject in higher organisms. First, we will consider enzyme induction and repression in adult organisms and then examine its possible relevance to cell differentiation during development.

Superficially, at least, many phenomena like enzyme induction and repression are seen to occur in higher organisms. That is, administration of certain substances to animals or plants can lead to increases or decreases in the levels of specific enzymes in particular tissues. In most organisms many such effects are known. The substances causing these changes in enzyme level are sometimes metabolically related to the enzymes changed, as are inducers in bacteria, or are sometimes not obviously related, like hormones. In some instances the level of the same enzyme can be changed by either of two such substances. For example, from Figure 13 you can see that the level of the enzyme tryptophan pyrrolase (an enzyme that splits the amino acid tryptophan) present in the liver of the rat can be 'induced' either by its substrate tryptophan or by the hormone hydrocortisone (HC).

If you compare Figure 13 with the data for the induction of β-galactosidase in *E.coli* (Fig. 7) one other point will emerge: the time scale for the effect is much longer. So what is happening in the first hour or so after the injection? Well, first of all we do not know from this experiment how long it takes the tryptophan or hydrocortisone to reach their sites of action and indeed where these sites of action are. It may not be the liver, where we observe the effect. Perhaps tryptophan has to affect the brain first, which then leads to nerve impulses which stimulate

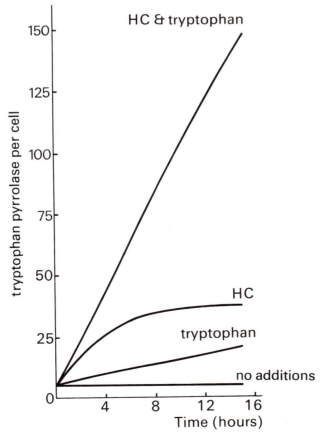

Figure 13 The induction of tryptophan pyrrolase.

the kidneys to produce a substance that induces tryptophan pyrrolase in the liver, and so on. This last suggestion though far-fetched serves to emphasize one point: when doing studies of this kind in higher organisms there are complications which do not exist when working on bacteria. There, for example, we add lactose to a homogeneous population of *E. coli* cells and can demonstrate easily that the lactose enters the cells within about 30 seconds. To establish similar facts about the substances administered to higher organisms, although possible, is much more difficult and time-consuming. This in essence is true for all studies on higher organisms, and for that reason our knowledge of the detailed mechanisms is much less than for bacteria.

Nevertheless, if in response to injection of a substrate, increases or decreases in the levels of certain enzymes are observed in a tissue in a higher organism, one can at least attempt to determine the subcellular events in that tissue that precede those changes. Here again the analysis of these events is more complicated than for bacteria as specific protein degradation is important in higher organisms. Thus, if protein molecules are being continually destroyed and replaced by newly synthesized ones, it is not enough to show that an increase in the level of an enzyme in response to, say, a hormone is paralleled by the incorporation of added radioactive amino acids into the enzyme protein, as it was for β-galactosidase (observation 4 on p. 24). The normal rate of synthesis of the enzyme may not be increased by the hormone, but it is possible that the rate of degradation of the enzyme is reduced. So any particular observed induction or repression must be carefully studied to see whether it is due to increased synthesis or reduced degradation. Relatively few exhaustive studies of this nature have been done and those there are have shown that, in principle, regulation of both the rates of protein synthesis and degradation are important mechanisms for controlling the level of specific enzymes in higher organisms. In fact, as those of you who have studied Biochemistry (S2-1) will know from Unit 6 of that Course, the induction of tryptophan pyrrolase in Figure 13 is an interesting example. It has been shown that hydrocortisone stimulates the rate of synthesis of the enzyme, while tryptophan stabilizes existing enzyme and slows down its degradation. That is why the combination of hydrocortisone plus tryptophan (Fig. 13) has a greater than additive effect. (For example, if hydrocortisone leads to an 8-fold increase in tryptophan pyrrolase and tryptophan makes it 4 times more stable this will give a 32-fold (8×4) increase in the level of enzyme.) However, for the induction of tryptophan pyrrolase by hydrocortisone and for many other inductions in higher organisms there is good evidence that a stimulation of specific protein synthesis is involved. It is now pertinent to ask whether the stimulation and its control operate *à la Jacob–Monod* or not.

2.5.1 Operons, regulator genes, repressors and mRNA in higher organisms

The Jacob–Monod hypothesis rests on the existence of adjacent structural genes (the Operon Concept) and specific regulator genes (e.g. the *i* gene) which code for repressors that control transcription of the structural genes. We shall now briefly examine the evidence for and against the existence of such entities in systems in higher organisms that exhibit enzyme induction or repression.

(a) Operons and regulator genes

The identification of operons (i.e. clusters of adjacent genes which contain information for metabolically related enzymes and which are controlled coordinately) in bacteria depends on the availability of genetic techniques that are good enough to definitely decide that any two genes are adjacent on the DNA and not just 'near each other'. Such techniques are not in general available in higher organisms. In some instances it can in fact be shown that the structural genes for enzymes that are induced or repressed together are definitely not all adjacent. Some such examples, where, in other words, more than one 'operon' exists for a particular set of induced enzymes, even occur in bacteria. The structural genes that code for the enzymes responsible for the synthesis of the amino acid arginine in *E. coli* are found in several separate clusters on the DNA, not all in one cluster.

> **ITQ 8** Can you suggest a simple modification of the Jacob-Monod hypothesis as shown in Figure 11 that will enable several operons to be under the same inductive or repressive controls?
>
> *Now check your answer with that on p 56.*

Regulator genes are operationally defined by isolation of mutants that are affected in the synthesis of specific enzymes whose structures are not affected (e.g. *i* gene mutants make β-galactosidase constitutively, but it is unaltered in structure). Some genes with these properties have been tentatively identified in higher organisms, but here again the evidence is less compelling than in bacteria.

(b) Repressors

The search for repressors in higher organisms has involved many groups of workers over the last 10 years or so. By definition almost, they have looked for proteins with the properties of inhibiting RNA synthesis from specific regions of the DNA. To do this, a cell-free system comprising DNA, RNA polymerase, nucleotide bases and metal ions has been mainly employed.

When incubated in appropriate buffers, such a system results in synthesis of RNA, this RNA being copied from the DNA provided. The problem is then to find proteins that inhibit this RNA synthesis. The main candidates as repressors were, for a number of years, a class of proteins called *histones*. Histones are proteins that contain a large amount of the positively charged amino acids, arginine and lysine. They are found in the nuclei of most cells in higher organisms where they are tightly bound to the DNA. If some isolated histones are added to an RNA-synthesizing cell-free system, as described above, they tend to inhibit RNA synthesis. This ability of histones to inhibit RNA synthesis and their normal association with DNA led to their being considered as repressors.

histones

> QUESTION Can you suggest which other characteristics histones should have as repressors? Would you expect histones to comprise a large number of different proteins or not, and if so why?
>
> ANSWER Any given repressor must be very specific for one particular operator region; that is, it must control the synthesis of a specific set of enzymes. As in any cell there are many enzymes subject to independent induction or repression, there must be many different repressors in that cell. So histones, if isolated from a cell *en masse*, should be a mixture of many different proteins.

Examination of histones show that in fact the histone fraction from any one cell does not comprise very many different proteins. Furthermore, careful analysis of some types of histones from two widely different species, the pea plant and the cow, show them to be very similar. As histones seem to lack specificity they are unlikely to be repressors in the Jacob–Monod sense. More recent candidates are a group of negatively charged proteins which are also found in association with DNA. Here, however, it is too early to say whether they are true repressors or not.

(c) mRNA

In certain tissues of some higher organisms where substrates or hormones cause enzyme induction, it is found that RNA synthesis increases in these tissues before any increase in enzyme. This suggests, that as for the induction of β-galactosidase, RNA synthesis is necessary for induction. This view is partly reinforced by the finding that substances that are known to inhibit all RNA synthesis also inhibit certain inductions. However, both these observations are unsatisfactory; the increased RNA synthesis does not mean that this RNA must be mRNA for the enzymes induced. Likewise, as the inhibitors stop all RNA synthesis they are very poisonous substances and sometimes the inhibition of enzyme induction may merely be a manifestation of a dying organism. Ideally, one would like to isolate the RNA whose synthesis is increased and identify it as mRNA for the enzymes induced, if indeed this is the mechanism involved. However, as we shall see, such an isolation is not easy! Other studies of mRNA in higher organisms have further complicated the issue. One of the factors on which the Jacob–Monod scheme for explaining induction and repression rests is that mRNA is unstable. Thus, inhibition of specific mRNA synthesis (by repressors) will inhibit further protein synthesis; the switch-off is rapid, as we have observed (2.4.2). However, examination of the half-lives of various mRNAs in higher organisms (the techniques involved are very indirect and will not be dealt with here) has revealed that they are frequently of the order of hours or even days, not minutes as in bacteria. Thus even if transcription of such mRNA was switched off, the existing mRNA could be translated to give protein for some considerable time (i.e. as long as the mRNA survived). This suggests that where rapid changes are seen in enzyme levels in higher organisms, these inductions and repressions may be due to the regulation of the rate of translation of stable mRNAs (alternative (b) p. 23) rather than regulation of their synthesis, unlike induction of β-galactosidase in *E. coli*.

By now you are probably aware that the whole question of the mechanisms by which the synthesis of specific proteins is controlled in higher organisms is in a confused state. *We suggest that you now summarize the main points of these Sections (2.5–2.5.1) and then compare your summary with that given below. Once again, if the summaries differ very widely, you should read these Sections again.*

Summary of Sections 2.5–2.5.1

1   Enzyme induction and repression occurs in higher organisms.

2   'Inducers' can be substrates or other substances such as hormones.

3   Studies are complicated by:
(a)   the problem of whether or not the administered substance directly affects the cells in question.
(b)   The fact that effects may be due to alterations in synthesis or degradation.

4   The Jacob–Monod hypothesis is not very well established for higher organisms as evidence for operons, regulator genes and repressors is very sketchy.

5   mRNA in higher organisms is often quite stable. This means that some controls may operate by regulating the rate of mRNA translation rather than its synthesis.

6   Some cases of enzyme induction and repression in higher organisms probably operate via alterations in mRNA synthesis, others by alterations in mRNA translation.

SAQ for Sections 2.5-2.5.1

SAQ 5 (Objective 7)   On injecting substance E into a rat there is a progressive increase in the level of an enzyme that destroys E, E-ase, in the liver. Other proteins do not alter in level. The time course of this increase is shown in Figure 14.

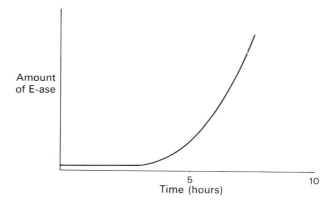

Figure 14

Which of the following statements are likely to be true and which false?
(a)   The delay in the increase in E-ase is due to the slowness of cell division in liver cells.
(b)   The increase in E-ase could, in principle, be demonstrated to be due to an increased rate of protein synthesis by injecting radioactive amino acids along with E and isolating E-ase seven hours later and showing it to be radioactive.
(c)   If the increase in E-ase is solely due to an inhibition of the rate of degradation, then the increase would not be expected to be altered by the addition of inhibitors of protein synthesis.
(d)   The increase in E-ase could be due to an increase in the rate of translation of mRNA for E-ase.
(e)   The increase in E-ase could be due to an increased level of ribosomes in the cell in response to E.

2.6   Summary of Unit 2

1   Cell differentiation involves an interaction between cells and their environment.

2   Cell differentiation does not involve a loss of DNA, that is, differentiated cells (like the zygote) are totipotent.

3   Different cell types arise due to differences in the proteins that they contain.

4   Cell differentiation must involve a switching on and off of genes.

5   In bacteria, control of protein synthesis is the major way of regulating levels of particular proteins.

6   Induction and repression in bacteria operate by regulating the rate of transcription by means of specific repressors.

7   In higher organisms both control of protein synthesis and degradation are important means of regulating the levels of particular proteins.

8   Control of protein synthesis in higher organisms in some instances probably involves regulation of the rate of transcription; in others it probably involves regulation of the rate of translation of mRNA.

Unit 3

3.1 Introduction

In Unit 2 we established that cell differentiation involved a controlled change in the protein composition of the cells. This led us to consider the mechanisms by which this might occur. The studies that we examined were all done either on bacteria or already differentiated cells in higher organisms. These studies established that in higher organisms changes in the protein composition of cells could arise as a result of specifically controlled protein synthesis or degradation. But all these studies involved changes in response to short-term environmental changes, the sorts of fluctuations that might occur daily: a change in the output of the hormone thyroxine from the thyroid gland (S100, Unit 18), and so on. Are the findings from these sorts of changes, reproduced in the laboratory, relevant to the longer-term, apparently stable changes that occur during cell differentiation? If they are, then hypothesis II in Figure 6 (p. 19) seems to fit the data best, as it involves differences in both protein synthesis and degradation between different cell types. In this Unit we are going to examine, in the light of what we know from Unit 2, the relevance of some aspects of this hypothesis to cell differentiation.

In Unit 2 we biased our discussion towards the two major mechanisms by which specific protein synthesis can be controlled—the control of gene transcription (as in the case of β-galactosidase) and control of translation (as seems to occur sometimes in higher organisms on stable mRNA). We neglected a closer examination of the detailed mechanisms for specific protein degradation. There are two reasons for this:

1 Very little is known about the mechanisms for specific protein degradation.

2 It seems to us inherently more likely that the major differences between different cells would depend on differences between what they make (i.e. protein synthesis) rather than on all cell types making everything and some types breaking down some proteins, while others break down other proteins.

The extreme version of the synthesis and degradation scheme which is represented by Hypothesis III in Figure 6, seems wasteful. So, these two reasons are, if you like, our excuses for concentrating in the rest of this Unit on the evidence for control of specific protein synthesis as an explanation for cell differentiation, and on how these controls seem to be exercised.

3.2 Differential gene transcription

The control of synthesis of many proteins, notably those in bacteria, seems to depend on controlling specific gene transcription. The best understood case is of course that of β-galactosidase, and here the mechanisms seem to operate as suggested by Jacob and Monod. However, there are in principle other ways of achieving some control of specific gene transcription without involving regulator genes, repressors and operator regions; one such way we will discuss in Section 3.5.1. The degree of detail of the study of cell differentiation itself in no way permits one to infer the existence of operons and repressors, as you saw from Section 2.5.1. Nevertheless, there have been some interesting studies that do suggest that the control of gene transcription, irrespective of the involvement or not of repressors, etc., has an important role in cell differentiation. That is, cell differentiation may involve differential gene transcription, so that in different types of cells different sets of genes are transcribed.

QUESTION If differential gene transcription is involved in cell differentiation, what would you predict about the types of mRNAs contained in different types of differentiated cells?

ANSWER Different cell types should contain different mixtures of mRNAs as they would transcribe different genes. Thus any one cell type should contain some mRNAs unique to that cell type (i.e. mRNAs corresponding to proteins unique to that cell type). It should also contain some mRNAs in common with other cell types, as some types of protein are common to many cell types in any organism.

We can represent the situation as in Figure 15 where, for the sake of argument, we consider the predictions based on two hypotheses about the types of mRNA contained in, say, rat liver and kidney cells. In hypothesis I (Fig. 15) differential gene transcription is postulated to occur in differentiated cells; in hypothesis II it is not and therefore both cell types have the same content of types of mRNAs. In hypothesis II presumably some mechanism other than differential gene transcription would have to be invoked to explain the observed differences in protein content (e.g. control of translation of these mRNAs).

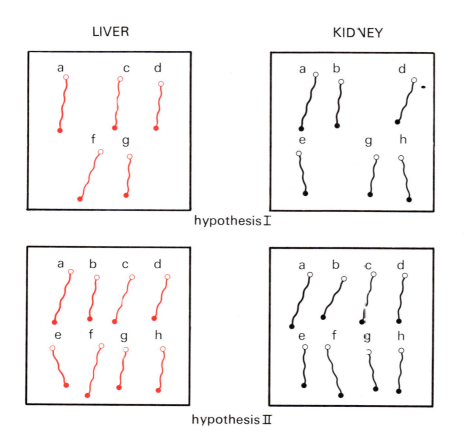

Figure 15 This shows schematically the possible contents of types of mRNA based on the two alternative hypotheses. The types of mRNA are indicated by the small letters (a–h), tissue of origin by the colour—liver (red), kidney (black). However, mRNA type a (mRNA$_a$) is the same substance whether from liver or kidney, etc.

To test whether the hypothesis of differential gene transcription (I in Fig. 15) has any validity or not, it merely remains to isolate the mRNAs from several different types of cell, separate all these mRNAs and identify them. On hypothesis I different types of cell would yield some differences in the types of mRNA that they contain; on hypothesis II, they would not.

In practical terms there are several difficulties—to name but a few:

1   It is not easy to isolate from cells RNA in an undamaged form.

2   It is very difficult to separate different mRNAs from each other.

3   It is difficult to 'identify' any one mRNA.

The last point is very pertinent. It essentially demands a means of measuring, (an assay for) the amount of a particular mRNA. In principle, there are two ways of doing this depending on the two specificities of any given type of mRNA:

(a)   It produces a specific polypeptide chain.

(b)   It is a copy of a specific sequence of nucleotide bases in the DNA (a gene).

We will briefly consider methods based on these two specificities.

(a)   Specific protein synthesis

From cells, it is possible to isolate ribosomes, activating enzymes, tRNAs and other supernatant factors. When provided with ATP, amino acids and mRNA, such cell-free preparations can synthesize small amounts of polypeptide (S100, Unit 17, Appendix 1). In principle, the polypeptides produced will depend upon the mRNAs added. Therefore, taking some of a total mRNA preparation from one particular cell type, adding it to a cell-free system and examining the polypeptides produced in consequence, one could categorize the mRNA content of that cell type. Such a test should distinguish between the two hypotheses. In practice, it is impossible, as yet, to do so. The reasons are numerous—to give but two:

1 The cell-free system is very inefficient and only minute amounts of polypeptides are made, often incompletely so. It is also doubtful whether any such system translates all the mRNAs present equally well.

2 Any one cell type (e.g. liver or kidney) would contain so many different types of mRNAs (probably hundreds) that the problem of separating the resulting polypeptides and identifying and distinguishing between them presents a fantastic challenge, as yet unmet.

QUESTION Assuming the polypeptide produced by translating mRNA$_a$ is a′, by translating mRNA$_b$ is b′, etc., tick the boxes below to indicate those polypeptides that would be produced when (1) liver mRNA, (2) kidney mRNA, was added to a ribosome cell-free preparation depending on whether hypothesis I or hypothesis II was correct.

Polypeptides made

Hypothesis		a′	b′	c′	d′	e′	f′	g′	h′
I	Liver								
	Kidney								
II	Liver								
	Kidney								

ANSWER

	a′	b′	c′	d′	e′	f′	g′	h′
I	✓		✓	✓		✓	✓	
	✓	✓		✓	✓		✓	✓
II	✓	✓	✓	✓	✓	✓	✓	✓
	✓	✓	✓	✓	✓	✓	✓	✓

(b) DNA–mRNA hybridization

The other assay system, which has been used with some success, depends on the fact that any given type of mRNA has a sequence of nucleotide bases that is complementary to one strand of a particular region of DNA (from which it was transcribed) from the same species of organism. This is because that region of DNA, which differs for each different type of mRNA, is the gene from which that mRNA was originally transcribed. This is shown schematically in Figure 16.

Figure 16 The DNA double helix is shown by solid lines. The mRNAs (types t, w, z) are shown as wiggly lines and as being complementary to the lower strand of the DNA. mRNA$_t$ was transcribed from, and is therefore complementary to, region (gene) T, mRNA$_w$ to W, etc.

This property of mRNAs can be used as the basis of an assay system as follows: As you know from Unit 17 of S100, the two strands of the DNA double helix are held together by hydrogen bonds. If a solution of DNA is heated to about 100 °C these hydrogen bonds are ruptured and the two strands come apart. If the solution is cooled rapidly, they remain apart, thus giving one a preparation of single-stranded DNA, that is, a preparation containing equal amounts of the two separated strands. Under appropriate conditions, single-stranded DNA can be made to form double-stranded regions with mRNAs. For example, on mixing single-stranded DNA with a mixture of, say, three types of mRNA (t, w and z), a DNA–mRNA hybrid molecule will be formed as shown in Figure 17.

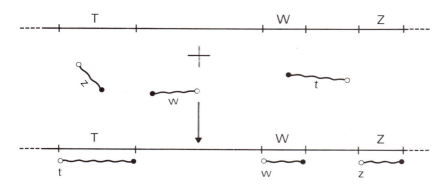

Figure 17

For any one molecule of DNA only one molecule of a particular mRNA (and hence complementary to a particular gene) could be bound: for example, in Figure 17 only one molecule of mRNA$_w$ can bind to W. So if a large amount of a preparation of mRNA (i.e. there are many molecules of each type of mRNA that the particular mixture contains) is added to a relatively small amount of single-stranded DNA (i.e. only a few molecules of each of the two strands of the DNA are present), all the regions of the DNA that are complementary to the types of mRNA present will be bound by mRNA, that is, the DNA will be saturated by that preparation of mRNA. Since the mRNA is added in a large excess, there will be more of each type of mRNA present than there are complementary regions on the DNA molecules to bind them. So some molecules of mRNA will remain unbound to DNA and hence not in the DNA–mRNA hybrid molecules. Indeed, individual molecules of the same type of mRNA will 'compete' for binding to those same regions on the DNA to which they are specific. This competition between like types of mRNA is used to compare the similarities and differences between the mRNAs of different cell types.

> **ITQ 9** Using, as we have in Figure 15, red for mRNA from liver and black for mRNA from kidney, sketch a diagram to indicate what would happen if mixtures of both rat liver and kidney mRNA preparations were added to a small amount of rat DNA, depending on whether hypothesis I or hypothesis II is correct. Where would the mRNAs compete and where not? Assume the DNA contains regions A–H (where A is complementary to mRNA$_a$ etc.).
>
> *Now check your answer with that on p. 56.*

As you can see from ITQ 9, hypothesis I predicts that competition amongst the mRNA molecules would only occur for genes A, D and G; on hypothesis II competition would occur at all genes (A–H). In order to make the degree of competition that actually occurs measurable experimentally one needs a way of distinguishing whether mRNAs in the mixed preparation come from liver or kidney (i.e. something analogous to red and black as used in our diagram). This is achieved by radioactively labelling the mRNA preparation from one cell type and not the other. Say, for example, the mRNA preparation from the liver is radioactive and that from kidney is not. This allows one to measure the amount of radioactivity (i.e. liver mRNA) bound to the DNA in the presence of increasing amounts of non-radioactive (i.e. kidney) mRNA. If competition is very high because the cell types have very many mRNA types in common, then the non-radioactive mRNA could eventually (at high enough amounts) almost completely 'compete-out' (i.e. exclude) any radioactive mRNA from the DNA–mRNA hybrid molecules. If competition is low, as some mRNA types are unique to each cell type, then not all of the radioactive mRNA can ever be removed from the DNA–mRNA hybrid molecules. Look again at Figure 34 and consider that red is equivalent to radioactivity, black is equivalent to non-radioactivity, and you will see what this means in terms of measurable radioactivity. Thus, the two hypotheses (I and II) theoretically predict the results shown in Figure 18 for such an experiment.

When such an experiment is actually done, the results obtained are like those shown below in Figure 19. The addition of increasing amounts of non-radioactive liver mRNA acts as a control to show that non-radioactive liver mRNA is exactly like radioactive liver mRNA and thus competes completely.

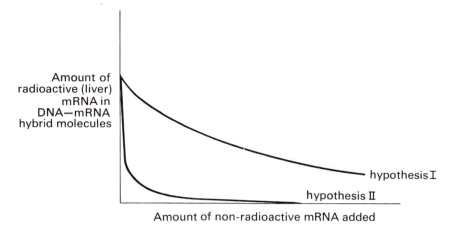

Figure 18

From Figure 19 it appears that the actual results fit the predictions made on hypothesis I.

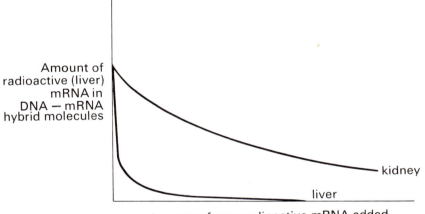

Figure 19 Large amounts of radioactive mRNA preparation from liver were added to small amounts of rat liver DNA to saturate it. This was done in the presence of increasing amounts of non-radioactive liver or kidney mRNA.

QUESTION Must the differences seen between the RNA preparations from kidney and liver (Fig. 19) be due to differences between the mRNA types in these two organs?

ANSWER Not necessarily, as in addition to mRNA, ribosomal RNA and tRNA would also be expected to hybridize to DNA. It is generally assumed, however, that ribosomal RNA and the tRNAs will be the same in the different cell types within any particular organism, and therefore it is probably reasonable to assume that at least some of the differences noted are indeed due to real differences between the mRNA types present in different tissues.

This is consistent with the hypothesis that, at least in some part, the differences in the protein composition that occur during cell differentiation can be explained by differential gene transcription.

One could perhaps object to the above conclusion on the grounds that the differences observed in mRNA content were found between already fully differentiated cells, kidney and liver. Perhaps these differences arise as an end-product of cell differentiation not as a cause. They could arise as a consequence of the differences between cell types in their responses to what we earlier called short-term fluctuations. There are, however, some other experiments bearing on this issue.

It can, for example, be shown that the liver cells of foetal mice differ somewhat in the content of mRNAs from the liver cells of adult mice. So development has involved changes in the mRNA content. However, even foetal liver cells are quite differentiated from other cell types. Some other studies have been done on much earlier developmental stages. This time the animal used was an echinoderm,

the sea-urchin. The ability of non-radioactive preparations of mRNAs from various stages of the developing sea-urchin to compete for DNA with a radioactively labelled mRNA preparation from the 'prism' stage of development was examined. The results, given in Figure 20, show that there are changes in the types of mRNA present in the cells during the early stages of sea-urchin development.

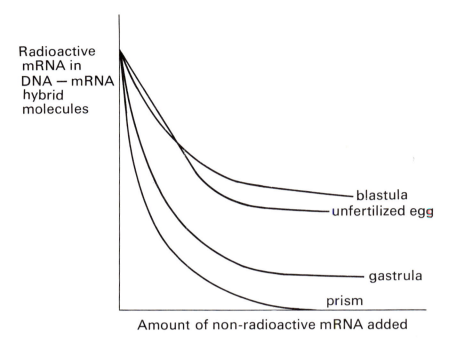

Figure 20 The radioactively labelled mRNA used was from prism. It was mixed with varying amounts of non-radioactive mRNA from one or other of the embryonic stages shown and the degree of competition with each type of mRNA measured. The order of the early embryonic stages in the sea-urchin is: unfertilized egg → fertilized egg → blastula → gastrula → prism.

It should, of course, be noted that in these experiments on sea-urchin, the mRNA preparations are derived from whole embryos. It would be difficult to obtain enough cells of individual types to do the experiment on.

Ideally, one would like a progressive study of the mRNAs present in the gradually changing cell types from egg to adult. This has not yet been achieved. Nevertheless, the studies done on individual types of fully differentiated cell (as in Fig. 19) coupled with those on early sea-urchin development (as in Fig. 20) do suggest very strongly that cell differentiation is accompanied by changes in the types of mRNA present in the different cell types.

But do the differences found in the types of mRNA present in the different cell types arise by differential gene transcription as we have assumed? We should perhaps consider the possibility that at least some of these differences in fact arise by differential mRNA degradation. That is, in different cell types even when the same genes are transcribed the mRNAs thus produced, have different stabilities in the different cell types. So we might be witnessing the result of selective mRNA degradation. Indeed, there is some evidence to support the idea that the stability of the same type of mRNA can vary from one type of cell to another. Could all the differences we have discussed arise in this way? *The answer is no*, for as you will see from TV programme 3 there is independent evidence to show that *at least some* differential gene transcription actually does occur during cell differentiation.

We can conclude that different cell types do contain some differences in their mRNAs and that these differences probably arise by differential gene transcription, although some may arise by differential mRNA degradation. These differences in mRNA content can, of course, in themselves explain some of the differences in protein content between different cell types. After all a cell cannot make, say, protein x′ if it does not make mRNA$_x$. Nevertheless, even if a particular mRNA such as mRNA$_x$ is present in a cell, its translation to produce x′ may itself be controlled and this too, as you will see, could be an important mechanism contributing to the differences in protein composition that occur during cell differentiation.

3.3   Control of translation

Experiments on induction and repression of enzymes in higher organisms have suggested that sometimes these phenomena come about by controlling the rate of translation of specific mRNAs rather than by controlling the rate of transcription. In essence, this rests on the observations that some increases in enzyme synthesis can occur even under conditions where no RNA synthesis takes place. Presumably, in this instance, the mRNAs for those enzymes are stable and are already present in the cells where their rate of translation is stimulated by the inducers. We will now examine whether regulation of translation is relevant or not to cell differentiation. To do this, we will consider just one set of experiments, though several similar studies exist.

The system under consideration is one already mentioned, the sea-urchin. Following fertilization of a sea-urchin egg, there is a rapid increase in the rate of protein synthesis in the cell. The new high rate of protein synthesis is maintained as cell division proceeds but then drops. There is a second increase just before gastrulation in the embryo (see Fig. 21).

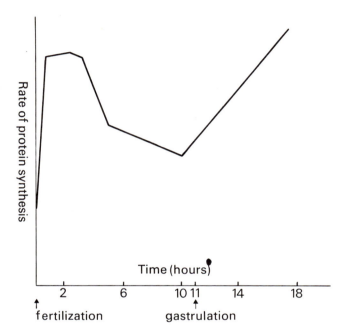

Figure 21 The rate of protein synthesis following fertilization in the sea-urchin embryo.

As well as a change in the *rate* of overall protein synthesis, there is also some evidence that there are changes in the *types* of proteins made. However, here again, as explained in the studies on sea-urchin mRNAs, individual cell types have not been examined. Do all these changes involve changed transcription or not?

A partial answer is obtained from the experiments with an antibiotic called actinomycin D. This substance is a powerful inhibitor of RNA synthesis, that is, it stops gene transcription. In its presence, fertilized sea-urchin eggs can develop as far as the blastula stage but do not gastrulate. This suggests that the formation of the blastula does not need new gene transcription but gastrulation does. Likewise, actinomycin D does not inhibit the increase in protein synthesis seen following fertilization but does inhibit that before gastrulation (see Fig. 22).

Thus, the increase in protein synthesis that occurs after fertilization does not involve new gene transcription. You already know that even the unfertilized egg contains some mRNA (see Fig. 20) and this is similar to that in the blastula (compare the curves in Fig. 20 for the unfertilized egg and the blastula). This mRNA is maternal, that is, it is present in the cytoplasm of the unfertilized egg. Presumably following fertilization there is a change in the rate of translation of such mRNA as is already present. This allows sufficient synthesis of the correct type of proteins to permit the embryo to proceed to blastulation. Since actinomycin D does block the pre-gastrulation increase in protein synthesis and also blocks gastrulation, it seems reasonable to conclude that gastrulation does require transcription of some 'new' (i.e. previously untranscribed) genes. This

maternal mRNA

again ties in with the appearance of different mRNAs in gastrulae (Fig. 20) from those in blastulae and an observed change in the types of protein in gastrulae also.

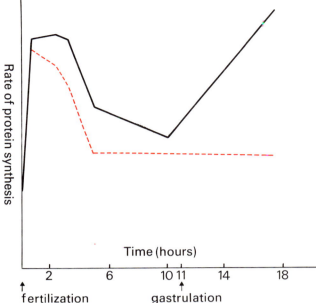

Figure 22 The effect of actinomycin D on protein synthesis following fertilization of the sea-urchin egg. (Red curve: plus actinomycin D.)

The above interpretations are in fact somewhat pat. Blastulae formed in the presence of actinomycin D are not entirely normal. They do probably contain a few minor differences in their spectrum of proteins. It can also be seen from Figure 20 that the mRNAs from blastulae and unfertilized eggs are not identical, implying that some mRNA synthesis occurs by the blastula stage. However, presumably none of these mRNAs synthesized after fertilization, nor any proteins synthesized from them, are vital to blastula formation as such, as the inhibition of synthesis of these mRNAs by actinomycin D does not prevent blastulation. Probably some of this mRNA is stored and translated at later stages, much in the same way as the maternal mRNA in the unfertilized egg is only translated after fertilization. However, gastrulation is inhibited by actinomycin D and presumably at this stage of development new RNA synthesis is necessary. This RNA probably includes new types of mRNA and ribosomal RNA (S100, Unit 17), as it has been shown that ribosomal RNA synthesis occurs between the blastula and gastrula stages.

These observations on the early development of the sea-urchin are still far from conclusive. They also differ in detail from similar studies on the timing of RNA synthesis during amphibian development. We can, however, reasonably conclude four important points:

1   The early changes in protein synthesis following fertilization largely depend on translational control of maternal mRNA.

2   Later changes depend to some extent on gene transcription to give new types of mRNA.

3   The translation of mRNA does not necessarily occur immediately after it is formed. That is, translational control exists (e.g. post-fertilization translation of maternal mRNA and post-blastula translation of mRNA formed during blastulation).

4   Control of gene transcription and mRNA translation are both important ways of controlling protein synthesis during the early development of embryos.

Summary and conclusions to Sections 3.2 and 3.3

1   Changes do occur in the protein content of cell types during cell differentiation.

2   There are accompanying changes in the types of mRNA present in the different cell types. At least some of these changes arise by differential gene transcription.

3   Some changes in protein synthesis during the early stages of development are independent of mRNA synthesis (i.e. gene transcription).

4   Mechanisms involving the control of protein synthesis at the level of gene transcription, and at the level of mRNA translation, play important roles during cell differentiation.

SAQs for Sections 3.1–3.3

SAQ 6 (Objectives 1 and 8) The diagram below shows the effect of adding increasing amounts of non-radioactive mRNA (from pig's heart or pig's liver) on the amount of radioactive heart mRNA that is found in hybridization with pig's heart DNA.

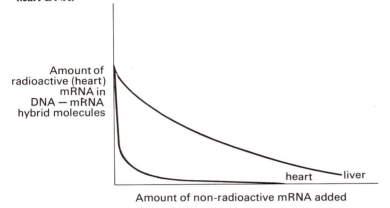

Figure 23

(a)   Does this show that liver and heart contain *some* of the same types of mRNAs or not?

(b)   What result would you expect if the experiment were repeated using DNA from pig's liver instead?

SAQ 7 (Objectives 8 and 10) If two cells contain some types of mRNA in common, does this mean that they also contain the corresponding proteins in common?

3.4   Nucleus and cytoplasm

The preceding Sections of this Unit have helped establish that the control of specific protein synthesis via control of transcription and translation can reasonably be supposed to occur during cell differentiation. This is only one part of the story. We argued in the introduction to Unit 2 that cell differentiation must depend both on the changes that occur within the cells and the 'signals' in the environment that trigger these changes. The study of the induction of β-galactosidase in *E. coli* has given a clear idea of what we mean by 'signals'; here the signal is lactose. But in higher organisms, where the changes that occur during cell differentiation are more stable, do they depend on a continuous signal (i.e. by analogy, does lactose always have to be present) or are the switching mechanisms harder to reverse? We cannot as yet answer such questions and it is difficult to say how valid an analogy 'lactose' is. In Units 4 and 5 we will consider what the signals from outside any one cell might be. However, there is one other type of signalling within a cell which was alluded to in our discussion of Gurdon's experiments (Section 2.2.1): that between nucleus and cytoplasm. It is this relationship that we will now consider, albeit briefly.

One of the striking differences between a bacterial cell and one of a higher organism is that the latter has a nuclear membrane and hence a distinct nucleus. Since the DNA (and hence transcription) is in the nucleus, while the ribosomes (and hence translation) are in the cytoplasm, the transport of mRNA from nucleus to cytoplasm may be yet another point of control of specific protein synthesis in higher organisms. But what is the relationship between a nucleus and the cytoplasm in which it exists? Is it merely a structural relationship or can each entity influence the activity of the other? The experiments of Gurdon (Section 2.2.1) provide a ready answer. The nucleus of a differentiated cell is totipotent yet it does not express its full potential in that cell, as the cell is by definition differentiated. However, when transplanted to another type of cytoplasm, that of an enucleated zygote, this nucleus changes in its behaviour and expresses its full genetic potential as would a zygote nucleus. Thus, it must be possible for the cytoplasm to influence the nucleus, the transfer of information is not just one way.

The information from nucleus to cytoplasm is in the form of mRNAs, the information in the reverse direction probably includes signals to produce certain mRNAs. This last statement is rather speculative but is borne out by some experiments by Gurdon.

Consider the following data:

1 If one examines cells in a frog blastula they can be seen to be actively synthesizing DNA, but not synthesizing RNA. However, if a nucleus from such a cell is transplanted into the cytoplasm of an unfertilized frog's egg, within a few hours the nucleus stops synthesizing DNA and starts to make RNA.

2 Likewise, nuclei in neurula cells (the neurula is a developmental stage in amphibia) synthesize RNA. If such nuclei are transplanted into the cytoplasm of fertilized eggs, within an hour they cease synthesizing RNA.

So these data do indeed suggest that the cytoplasm can modify transcription in the nucleus and that different cytoplasms have different effects.

These and other interesting experiments by Gurdon promise to tell us more about these nuclear-cytoplasmic interactions. However, Gurdon's experimental system has certain drawbacks. It involves only one species, *Xenopus laevis*, and it is technically very tricky.

Recently, there have been exciting advances in ways in which to place nuclei in novel cytoplasms and it is to these we now turn our attention.

3.4.1 Cell hybrids

Since the introduction of methods of culturing animal tissues (Section 2.2.1), it has periodically been noticed that two or more cells could occasionally fuse together to give multinucleate cells. In the early 1960s it was shown that the incidence of such cell fusion could be dramatically increased by certain viruses. If, before adding it to the cells, the virus is irradiated with ultraviolet light, its nucleic acid is damaged and it cannot now multiply when added to the cells. Nevertheless, such a non-reproducing virus can still cause cell fusion as it is probably the protein coat of the virus that has this property.

The cells produced could be of two types:

(a) cells that contain two or more nuclei. These cells do not undergo successful cell division.

(b) cells that contain initially two nuclei but then during mitosis these nuclei fuse to give two daughter-cells, each with a large nucleus which contains the chromosome complement of both parent nuclei. Such 'double-sized nucleus' cells can divide as such for many generations.

In 1965, Henry Harris and his collaborators at Oxford University demonstrated that it was possible, using viruses, to fuse together cells from different animal species to give multinucleate cells. Some examples are shown in Figure 24.

Figure 24 (a) Tetranucleate cell derived from two HeLa cells and two Ehrlich ascites cells. The nuclei being different in appearance are readily distinguishable. (The HeLa cell nuclei are labelled H, the Ehrlich cell nuclei E.) HeLa cells are tissue culture cells derived from a human tumour. Ehrlich cells are a form of mouse tumour cell.
(b) Binucleate cell derived from one HeLa cell and one hen erythrocyte (a red blood cell). The HeLa nucleus is again labelled H.

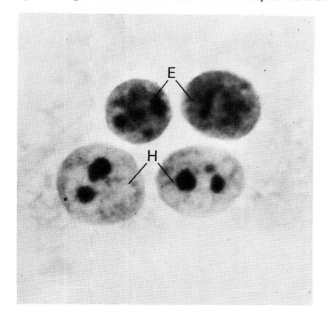
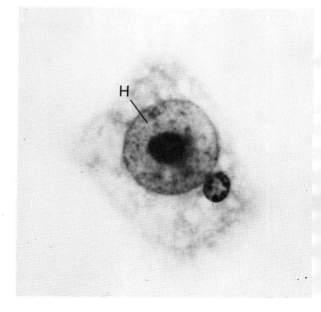

As with cells from a single species, binucleate cells derived from two different species could undergo nuclear fusion to give cells with a single large nucleus containing the chromosomes characteristic of both species. Such cells could divide for many generations. Reports of the work of the Oxford group, coupled with the idea that a Steward-type experiment could be done wherein a whole organism might one day be grown from an animal cell, resulted in some interesting speculations of varying degrees of scientific objectivity, to wit Figure 25.

Figure 25

Aside from these interesting ideas and the problems of whether a mouse-rat hybrid should be called a 'mat' or a 'rouse', the cell fusion experiments do allow one to investigate the inter-relationship between nucleus and cytoplasm.

For example, the nucleus of a rabbit macrophage cell synthesizes RNA but not DNA. However, if fused with a HeLa cell, the nucleus of which carries out both activities, in the resulting hybrid containing separate HeLa and macrophage nuclei, both HeLa and macrophage nuclei now synthesize RNA and DNA. The results from a whole series of similar experiments are summarized below.

Cell type	Synthesis of	
	RNA	DNA
HeLa	+	+
Rabbit macrophage	+	−
Rat lymphocyte	+	−
In hybrids:		
HeLa–HeLa	+ +	+ +
HeLa–macrophage	+ +	+ +
HeLa–lymphocyte	+ +	+ +
Macrophage–macrophage	+ +	− −
Macrophage–lymphocyte	+ +	− −

N.B. Each $+/-$ represents the activity of one nucleus.

Examine the table carefully. You will note that these results suggest that the regulation of nucleic acid synthesis is unilateral: whenever a cell that synthesizes a particular nucleic acid (DNA or RNA) is fused with one that does not, in the resulting hybrid both nuclei do so. This suggests that one nucleus activates the other, or that signals in the cytoplasm from the active cell cause this activation. But a cell hybrid as well as containing nuclei from each type of cell also contains both types of cytoplasm and so it is possible that each nucleus is still responding to signals from its own cytoplasm, these having been affected by the process of cell fusion. Fortunately, this possibility can be eliminated by examining hybrid cells in which one parent cell is a hen erythrocyte. This is because during the process of cell fusion the erythrocyte loses all its cytoplasm into the experimental medium. Thus, in a multinucleate hen erythrocyte–other cell hybrid, no erythrocyte-derived cytoplasm is present. Such a hybrid is shown in Figure 24b. The erythrocyte nucleus produces no RNA or DNA but when fused with active cells such as HeLa, here again both types of nuclei (HeLa and erythrocyte) in the hybrid produce these nucleic acids. Similar results can be obtained when hen erythrocytes are fused with cells from a wide variety of animals. In each case, only one type of cytoplasm is present, not that of the erythrocyte. Yet, in each instance, if the other cell is active in nucleic acid synthesis, the erythrocyte nucleus in the hybrid is activated. So the cytoplasmic signals that cause nuclear activation are not species specific. This is but one important conclusion that has come out of using this bizarre experimental system. We cannot deal with any more here but it is likely that this system will yield many interesting results and ideas about the reactions between nucleus and cytoplasm and their relevance to cell differentiation.*

3.5 Temporal control

We have now examined in some detail the changes that occur during cell differentiation and the subcellular mechanisms by which they come about. As you have seen, there are elaborate mechanisms to ensure that these changes are controlled. However, we have yet to discuss one control which we mentioned at the start of these Units (Section 2.1), temporal control.

This is a particularly important aspect of cell differentiation. If we can demonstrate that the control of timing of the changes involves a distinct sequence wherein one change is dependent on earlier ones, this answers in part the crucial question that we posed earlier of whether the observed changes are necessary components of cell differentiation or merely consequences of it and in particular, whether the observed changes in protein and mRNA contents are causal factors in cell differentiation (Section 2.3). Temporal control also bears on the question of the stability of the differentiated state. For example, at what stage does the state of a differentiating cell become irreversible? In other words, when does a particular cell or group of cells become determined (Unit 1). Though one can often answer such questions in higher organisms, it is very difficult because of the complexity of the changes occurring during development, the number of different cell types and the relatively long time spell over which the process stretches, to uncover the actual mechanisms underlying determination. For these and other reasons it is desirable to examine simpler systems which might prove 'models' for understanding the more complex ones. Rather in the way that induction of β-galactosidase in *E. coli* helps in the investigation of the control of protein synthesis in higher organisms, it assists one to ask the right questions. We shall now consider two such model systems, which pose interesting questions in themselves and give some insights into temporal control in higher organisms. The two systems are sporulation in bacteria and the developmental cycle of cellular slime moulds.

3.5.1 Bacterial sporulation

As you know, bacteria are unicellular organisms that reproduce asexually by simple cell division. Given favourable conditions of temperature, pH and food, bacteria grow and divide rapidly. Such dividing bacterial cells are said to be 'vegetative'. However in certain species of bacteria there are alternative life cycles: under adverse conditions, such as a lack of vital nutrients, a bacterial cell stops dividing and undergoes a series of biochemical and structural changes. These changes culminate in the production of a structure called a 'spore'. A spore   **spore** is a metabolically inert cell: it takes in no nutrients, hardly shows any metabolic activity at all and does not divide. Rather like a plant seed, it is inert. Like a plant seed, given favourable conditions a spore can 'germinate' to give ultimately a normal vegetative cell which then divides to give two cells, and so on. As well as being inert, a spore differs from a vegetative cell in one other important way: it is very resistant to adverse conditions such as high temperatures, lack of water, radiation, and lack of food. To a species of bacterium this alternative life cycle therefore has the evolutionary advantage of allowing the bacterial cell to survive, in spore form, adverse environmental conditions. Such is the resistance to heat and radiation of many bacterial spores that certain spore-forming pathogenic

* Those of you who are interested may like to read the account of his studies by Henry Harris, called *Cell Fusion* (Oxford University Press, 1970).

bacteria can present a problem in hospitals. This necessitates the sterilizing of equipment at temperatures well above those at which vegetative bacteria are killed. However, in certain conditions it is virtually impossible to sterilize something to the exclusion of all spores. It is highly probable that the man-made spacecraft that have landed on the moon, Mars and Venus, have contaminated those bodies with bacterial spores of earthly origin. Part of this resistance of spores is due to their possessing several outer layers or 'coats'. A fairly typical spore of the genus *Bacillus* is shown in Figure 26, along with a diagram summarizing the life cycles.

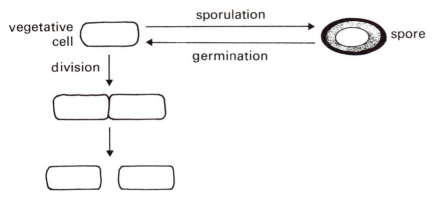

Figure 26 Alternative life cycles of *Bacillus*.

It should be noted that the process whereby a spore gives rise to a vegetative cell, which involves germination, is not just a simple reversal of the processes involved in sporulation. Both processes, sporulation and germination, are useful models for cell differentiation, and both have been studied as such. We shall just consider sporulation and, in particular, sporulation in just one organism, *Bacillus subtilis*, though much of what we will discuss is probably true for other species of bacteria too. In order to be brief, we will consider sporulation from just four standpoints:

1 Is the system a good model for cell differentiation in higher organisms?

2 Does it exhibit temporal control?

3 Is there a stage of determination?

4 How is the time sequence controlled?

1 *Sporulation—a model for cell differentiation?*

First let us consider the system of sporulation in *B. subtilis* in more detail. Placed in medium containing adequate nutrients, including sources of carbon and nitrogen (Section 2.4.1), *B. subtilis* grows and divides rapidly. If some such vegetative cells are transferred to a medium lacking carbon or nitrogen sources the cells cease to divide and give rise to spores within about 8 hours. During the sporulation several structural changes can be seen in the cells using a light or electron microscope. This allows us to identify several intermediate stages in the process of sporulation and these, as shown in Figure 27, are arbitrarily labelled from 0 to VII; stage 0 is the vegetative cell.

Associated with these structural changes there are several biochemical changes. A spore differs in several respects from the vegetative cell. For example spores contain large amounts of Ca^{2+} ions and a substance called dipicolinic acid (DPA). Presumably these changes require changes in protein composition to occur and indeed massive protein turnover (protein degradation and resynthesis) is found to occur during sporulation as 75–90 per cent of the cellular proteins in the spore are actually synthesized during sporulation. This is not to say that all these newly synthesized proteins are types unique to the spore, many of them are probably the same as those in the vegetative cells but some, such as the proteins contained in the spore coats and the enzymes needed to synthesize DPA, probably arise only during sporulation. Thus sporulation, like cell differentiation in higher organisms, seems to involve changes in protein composition that arise as a result of controlled protein degradation and synthesis. However we defined one other criterion for cell differentiation in higher organisms—choice. We have hitherto always considered evidence by comparing two cell types (e.g. liver and kidney) that are assumed to have arisen from a common type of 'ancestor' cell.

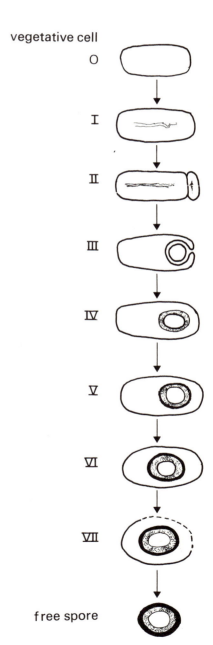

Figure 27 Stages in sporulation in *B. subtilis*.

47

This common ancestor divided to give two cells each of which, as it were, 'chose what to be' (Section 2.1, Fig. 2). In simple schematic terms

where A, B and C are cell types. In sporulation we appear to be just comparing a spore with its ancestor, a vegetative cell. However, as we pointed out in Unit 1 this is still an example of differentiation as sporulation does involve choice because a vegetative cell 'choses' to divide (i.e. remain a vegetative cell) or to form a spore, thus

(Note in this connection that the early stages of sporulation look like abnormal cell division.) So, in terms of choice too, sporulation seems to be a reasonable model for cell differentiation.

2 Temporal control

The process of sporulation involves several structural changes occurring in a defined sequence over a defined time period (Fig. 27). Associated biochemical changes can also be shown to occur with a defined timing (see Fig. 28).

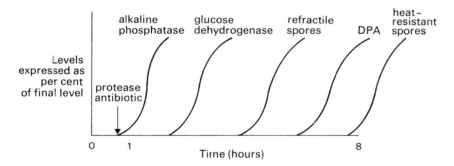

Figure 28 Schematic representation of the biochemical changes occurring during sporulation.

Time 0 is the time at which the vegetative cells are transferred to non-nutrient medium. At the time indicated by the arrow, the cells produce and secrete into the medium, an antibiotic and a protease (an enzyme that degrades protein). Alkaline phosphatase and glucose dehydrogenase are enzymes whose levels inside the cells increase as shown; refractile spores look different from vegetative cells under the light microscope. We do not understand the function of all these changes but they do serve as useful 'biochemical time markers' of the process.

However, given that sporulation involves several structural and biochemical changes occurring in a defined sequence, this does not prove the existence of temporal control. To do this one must demonstrate that the sequence is necessary for sporulation, that later events are dependent upon earlier ones. This is where genetics comes to the experimenters' aid.

It is possible to isolate many different mutants of *B. subtilis* that cannot sporulate. Such mutants are called *asporogenous*. Fortunately they fall into each of several classes, blocked at one or other stages of sporulation. There are, for example, asporogenous cells which when placed in non-nutrient medium undergo all structural and biochemical changes up to stage II, or others that go up to stage III (Fig. 27), and so on. What is significant is (a) a particular asporogenous mutant will only go through the changes characteristic of the stages before that at which it is blocked and (b) these changes are both structural and biochemical.

ITQ 10 Does this tell us anything about whether temporal control exists in this system or not, and if so what?

3 Determination

The process of sporulation starts when nutrients are removed from the vegetative cells. That is, the cells are committed (Unit 1) at that time. Once started, can the

48

process be reversed by merely adding nutrients back, or once started are the cells determined so that they become spores? In bacterial induction, as exemplified by the β-galactosidase system, we know that the process is easily halted by reversing the signal, that is, by removing lactose.

The experiments are relatively easy. Vegetative cells are transferred to non-nutrient medium and after differing periods of time some samples of these cells are transferred to nutrient medium. All samples are then left for some hours and examined to see if the cells have become spores or reverted to become vegetative cells. Some such experiments were done by Mandelstam and his co-workers. As nutrient they used a medium containing a mixture of amino acids, and as a measure of spores formed they counted the number of refractile spores that could be seen under the light microscope. Figure 29 shows the proportion of cells that reach to the stage of being refractile spores depending on the time at which nutrient is added back.

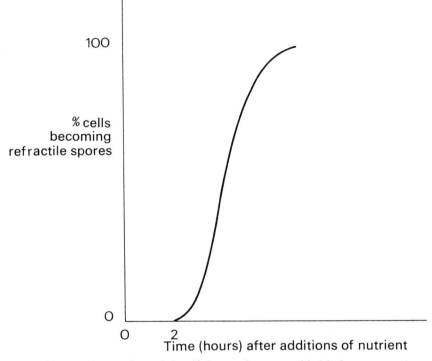

Figure 29 Schematic representation of the proportion of cells becoming spores as a function of the time at which nutrient is added to the medium.

Thus this experiment showed that if the nutrient was added before two to three hours all the cells would revert to being fully vegetative, however cells left longer than this were determined as refractile spores. Thus about three hours is the determination time for producing refractile spores. Mandelstam and his colleagues then repeated this experiment but instead of using refractility as a measure of sporulation they used various of the other changes, such as the production of alkaline phosphatase or DPA. They observed once again that there were determination times for these events but they differed for each event—the earlier the event the earlier the determination time. Here again the determination time was the time at which addition of nutrient was too late to stop the event occurring. So the determination time for the production of alkaline phosphatase, an earlier event (Fig. 28), is before that for production of refractile spores and that determination time before the determination time for DPA production, and so on. So, rather than there being a single determination time for the whole sporulation sequence, there appeared to be a series of sequential determination times for each of the events leading to spore formation. Obviously if we can unravel how determination comes about in chemical terms we can start to understand how temporal control operates. This we will now examine.

4   The control of timing

The determination time for any of the events occurring during sporulation is usually about an hour before that event manifests itself. That is, the time at which spores are determined to becoming refractile is about one hour before refractility is observed, and so on. So what is it that occurs at a determination time that fixes what happens one hour later?

A clue is given by some of the experiments carried out by Sterlini and Mandelstam using actinomycin D.

If actinomycin D, which as you know is an inhibitor of RNA synthesis (Section 3.3) is added to some cells at the beginning of sporulation (i.e. time 0), it prevents any sporulation. This is not really surprising, as from what we know about protein synthesis in bacteria and the changes that occur during sporulation, we would expect that the gene transcription is necessary for sporulation. However, if the experiment is repeated, but actinomycin D is added to samples of cells in non-nutrient medium at later and later stages of sporulation, it is found that the actinomycin D fails to inhibit sporulation entirely. Indeed it produces a set of results remarkably like those obtained by adding nutrient amino acids to sporulating cells. For comparison, in Figure 30 we present the effect of adding amino acids or actinomycin D to cells at various times after the removal of nutrient on the number of refractile spores those cells go on to produce.

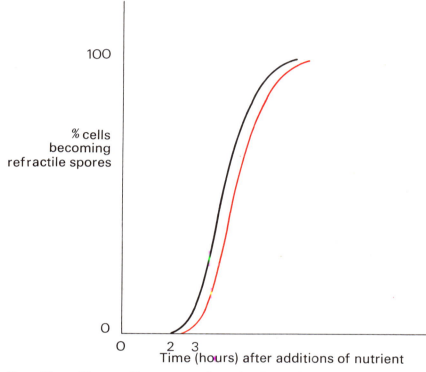

Figure 30 Schematic representation of the effect of actinomycin D (red line) and amino acids (black line) on sporulation.

From Figure 30 you will note two points: the shape of the two curves is very similar and there is a 20-minute gap between the curves; the formation of refractile spores becomes insensitive to actinomycin D within 20 minutes of the determination time.

This relationship between the curve for determination and that for the effects of actinomycin D is borne out for all the determination times (alkaline phosphatase, etc.) examined. It suggests that at each stage of determination there is a step that is inhibited by actinomycin D, presumably a synthesis of one or more mRNAs specific to that determination time and that within 20 minutes there is sufficient of this synthesis, and hence the sporulation event is unaffected by actinomycin D. Most probably the mRNAs synthesized at any particular determination time are those needed to synthesize the proteins necessary for the next sporulation event to occur. However the next event does not occur for an hour after the determination time whereas the actinomycin D experiments show that the synthesis of the mRNAs necessary for the event takes only 20 minutes.

ITQ 11 What does this suggest?

We can now suggest three important conclusions about the timing of the sporulation.

1 Any particular determination time involves a synthesis of a set of specific mRNAs.

2 Timing of sporulation involves timing of sequential transcription of different genes.

3 The gap of 40 minutes between the completion of synthesis of a set of specific mRNAs and their expression as a sporulation event suggests that some control of mRNA translation also occurs during sporulation.

So we are now in a position to rephrase part of the problem: how does a cell control the sequence in which different genes, all relevant to sporulation, are transcribed? No one yet knows the answer. There are however several interesting hypotheses, three of which we now consider.

Hypothesis A: Sequential transcription of sequential genes

All the genes of *B. subtilis* are contained along one molecule of DNA. The simplest hypothesis to explain the sequence of transcription of the genes involved in sporulation is to postulate that all the genes involved in any one stage of sporulation are clustered together in a block, and that all these individual blocks of genes are in one region of the DNA molecule and that they are arranged in the order in which they need to be transcribed. A molecule of RNA polymerase goes along this region transcribing the blocks of genes in sequence thus giving the timing at which these genes are expressed during sporulation (see Fig. 31).

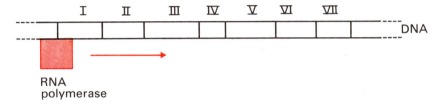

RNA polymerase

Figure 31 The Roman numerals correspond to blocks of genes. Each block is that relevant to the changes that occur during one stage of sporulation (the gene block numbers correspond to the stage numbers). Each block comprises all the genes relevant to that stage.

This hypothesis has been tested by examining genetically the various mutants blocked at different stages in sporulation. Genetic mapping techniques have shown that the various mutant genes are scattered throughout the DNA and there is not even any clustering into blocks of different genes relevant to the same stage. Thus the simple hypothesis outlined above is incorrect.

Hypothesis B: Alterations in RNA polymerase

To begin transcribing a particular gene, a molecule of RNA polymerase must attach to that DNA at the start of that gene. Such starting points must be 'recognized' by the RNA polymerase and are termed promoters (Section 2.4.2). As with an enzyme-substrate reaction this recognition depends on complementary specificities—the sequence of nucleotide bases in the promoter and the recognition site of the RNA polymerase.

It is possible to purify RNA polymerase from *B. subtilis*. Given the appropriate nucleotides, a DNA template and favourable ionic conditions, this enzyme can copy the DNA to give mRNAs. It has been shown by using DNA from more than one organism that RNA polymerase prepared from vegetative cells differs from RNA polymerase from sporulating cells in its capacity to transcribe these DNA preparations. This suggests that the ability of the RNA polymerase to transcribe particular genes changes during sporulation. This change has been tied-down to an alteration in one of the polypeptide chains that comprise RNA polymerase. So we could postulate that during sporulation the RNA polymerase was altered so that it could transcribe certain genes and not others. This could explain the changes in the genes transcribed at different times, but in order to explain the existence of several determination times several such changes would have to occur. This could not in itself explain the time sequence observed, as we would then be left to explain what controlled the order in which the different changes in the polymerase occurred.

Hypothesis C: Sequential induction

This hypothesis depends upon the idea that the first change produces something which then induces the enzymes necessary for the second change which then produces something which induces the enzymes for the third, and so on. This, like hypothesis B, can as yet be neither confirmed nor ruled out.

51

Summary and conclusions to Section 3.5.1

1 Bacterial sporulation involves a sequence of biochemical and structural changes.

2 The sequence is under temporal control as later events are dependent upon earlier ones.

3 The changes are similar to those occurring in higher organisms as they involve changes in protein synthesis and degradation. These changes in protein synthesis seem to be controlled by both the control of transcription and translation.

4 The temporal controls involve a series of controlled transcriptions—a series of determinations.

5 The temporal controls do not seem to involve a simple topological ordering of genes of the DNA, but may involve changes in the specificity of the RNA polymerase and sequential enzyme inductions.

In conclusion, you can see that the study of bacterial sporulation presents both a challenging and interesting problem in itself and *an instructive model for cell differentiation in higher organisms*. Of course, there are other aspects of the system we have not touched on, many of them very important. For example, what is the relationship between the vegetative cell cycle and sporulation? What causes the initial commitment towards sporulation? Ideas about these fundamental questions are only just beginning to emerge clearly. We cannot deal with them here but we recommend those of you who are interested to look out for the third level Course on Biochemistry and Molecular Biology (S322)*.

SAQ for Sections 3.5–3.5.1

SAQ 8 (Objectives 1 and 13) Which of the following statements are true and which are false?

(a) The sequential appearance of several enzymes during sporulation in bacteria proves that sporulation is controlled by sequential transcription of clustered genes.

(b) Not all of the types of protein found in spores differ from those found in vegetative cells.

(c) Changes in the structure and specificity of RNA polymerase during sporulation can in themselves explain the control of the time sequence.

(d) During sporulation there must be a marked change in the actual genetic information in the cell.

(e) The process of germination is controlled by a simple reversal of the events that lead to sporulation.

3.5.2 The cellular slime moulds

We have argued, we hope reasonably convincingly, that sporulation in bacteria does serve as a good model for cell differentiation in higher organisms. However, one aspect of this system as a model is we think a little shaky. Is a choice of the type

really the same as

We do not know. However, much of the justification of using sporulation in bacteria to study ultimately the mechanism of this choice hinges on whether they are similar or not. Ideally, what we would like is an organism that is

---

* As yet no final date has been set for the presentation of this Course.

experimentally simple enough to handle like a bacterium, but biologically complex enough to involve an

choice. One such system may well be the cellular slime mould where one type of cell (originally an amoeboid cell) gives rise to either a spore or a stalk cell. It is, of course, part of your home experiment to examine the life cycle of a particular slime mould, *Dictyostelium discoideum*. We leave it to your experiment plus the notes we give you and the associated TV demonstrations to deal with this problem.

Conclusions to Units 2 and 3

We have taken you at a rather hectic pace through a large number of experimental systems and conceptual ideas. We have tried to draw a thread through all our arguments and we hope you have maintained this thread and can see where the experiments are related to it. We decided on a continuous thread in our arguments to help make the mass of systems and ideas and experiments come together. This of course has consequences: we have restricted our discussion to those matters directly related to the theme and hereby we may have missed many crucial ideas and data. In all honesty whether we have or not we don't know. We have just presented some of the recent data in this rapidly expanding field and tried to indicate the lines along which future research is likely to go. But there are nagging doubts: Should we have considered control of DNA synthesis for example? After all, cells do divide during development and they need to synthesize DNA to do this. Does this bear on cell differentiation or not? Well, opinions differ. Perhaps that last sentence sums up much of the field of study of cell differentiation—a fact that makes the whole subject often very frustrating to study but, we hope, always very interesting.

But where does cell differentiation fit into the whole scheme of development? Even if we could list all the changes that occur to all the cell types in any organism, would this in itself allow us to totally describe or predict the form that the organism takes? Consider, as an example, a human hand and foot. Both contain skin, bone, tendon, blood, muscle, cartilage, and so on. They contain identical types of cells. Yet we can clearly distinguish a hand from a foot and they also occur at different extremities of the body. So just knowing the changes that occur to individual cells as they undergo differentiation is not enough. Knowing the interactions between these cells is also vital to any understanding of development. As Lewis Wolpert, a developmental biologist, says: 'How do you make a hand?' What are the changes, signals and genetic information? It is this problem that we tackle in the rest of this Course.

Answers to pre-Unit assessment test

N.B. All references in brackets are to S100.

1 True (15.3).

2 False. Ribosomes are where proteins are synthesized, ATP is mainly synthesized in the mitochondria (17.9).

3 False. Each strand of the DNA is a polynucleotide chain (17.2).

4 True (17.6).

5 False. This process is called translation (17.7).

6 False. A clone consists of genetically identical organisms (19.2).

7 False. Thyroxine is a hormone produced by the thyroid gland (18.3).

8 True (17.4).

9 False (19.3).

10 True (17.10).

11 (a), (b), (e), (f).

12 mRNA, activating enzymes (17.8).

13 Ionizing radiation, temperature, and certain chemicals are some examples of things that can cause mutations. (19.2).

14 True (19.2).

15 False. There are many different types of cells in a multicellular organism (14.4).

16 True (19.5).

17 True (19.5).

18 True (19.2).

If you were wrong more than 8 times out of 18, you should read again the appropriate sections of the Units in S100 before continuing with the text.

ITQ answers and comments

ITQ 1 (Objective 2) During differentiation some plant cells, such as xylem cells, lose their nuclei. As the genetic information, the DNA, is contained in the nucleus (S100, Unit 17) such cells cannot be totipotent.

Another problem with the Steward experiment is that, as you know, plants contain undifferentiated regions of growth (so-called meristems) throughout their lives. It is therefore important to ensure that the cells which in culture give rise to embryoids and then whole plants are not derived from meristematic tissue and hence not differentiated to start with. Careful repetition of these experiments on a variety of plants has made it likely that at least some cells which are definitely differentiated and not meristematic can give rise to plants and are hence totipotent.

ITQ 2 (Objective 2) As an embryo reaches later and later stages of development the more differentiated its cells would be and, therefore, if scheme A is correct, the more limited their genetic information. Even if zygote cytoplasm could act as 'coconut milk', nuclei from differentiated cells could not replace the totipotent zygote nucleus. So, the later the stage of development of the embryo from which the donor cell is isolated, the less it could support development in the 'nuclear-transplant egg'.

ITQ 3 (Objectives 1 and 4) A is mRNA, B is ribosomes, C is a polysome, D is polypeptides.
1 is transcription. Nucleotide bases are needed to help form mRNA (A).
2 is translation. Amino acids, activating enzymes and tRNA are needed to help form aminoacyl tRNAs which then, along with mRNA and ribosomes, are involved in translation.

3 is protein folding. This occurs spontaneously. One or more polypeptide chains released from the ribosome fold to give specific proteins.

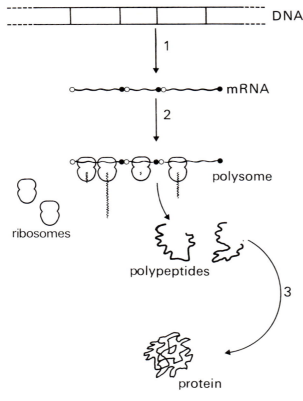

Figure 32

Substances other than those listed above are also needed. The overall process by which any one protein is synthesized may be summarized as follows:

1 *Transcription*

A gene is copied to give an mRNA. That is, the sequence of bases of one strand of one region of the DNA double helix is copied to give a complementary strand of RNA. This requires a divalent metal ion such as Mg^{2+} plus the four bases found in RNA in their triphosphate form: ATP, UTP, CTP, GTP where A is adenosine, U is uridine, C is cytidine, G is guanosine. The copying is catalyzed by an enzyme, DNA-dependent RNA polymerase (RNA polymerase for short), a molecule of which attaches to the beginning of each gene to be copied.

2 *Translation*

One or more ribosomes attach to each molecule of mRNA. The translation of the codons in this mRNA requires movement of the ribosomes along the mRNA and aminoacyl tRNAs, that is, molecules of each of the 20 amino acids attached to their specific tRNAs. The movement and joining of the amino acids to give a polypeptide chain also requires GTP. The joining of a particular amino acid to its correct tRNA requires ATP and an activating enzyme that is specific to that amino acid and that tRNA.

3 *Folding*

Completed polypeptide chains are released from the ribosomes whereupon they fold up spontaneously to give the specific three-dimensional shape characteristic of the protein in question. A protein can comprise one or more polypeptide chains, of one or more types.

ITQ 4 (Objective 4) 1 The specific genes (i.e. sections of DNA) which provide the information for the sequence of amino acids in the polypeptides.

2 The mRNAs copied from those specific genes.

3 The specific polypeptides that fold up to give the specific proteins.

ITQ 5 (Objective 7) (a) and (b) are still possible as 1-4 have little bearing on them. (c) is ruled out, since point 2 shows that non-substrates can be inducers (see Fig. 9) and points 3 and 4 show that induction does not depend merely on pre-formed polypeptides.

ITQ 6 (Objective 7) One still cannot decide between them, since point 6 is consistent with (a) or (b). It is interesting to consider the role of the *i* gene though. It could perhaps control the rate of β-galactosidase mRNA transcription ((a)) or its translation ((b)), but how?

Turn back to p. 25 and check.

ITQ 7 (Objective 7) No. We also cannot as yet on this data decide between (a) or (b). Neither could Jacob and Monod in 1961. They plumped for (a) but equally well mRNA synthesis could continue all the time and repressor could inhibit translation by binding to the end of the mRNA, perhaps preventing ribosomes attaching.

ITQ 8 (Objective 6) One solution is that each of the separate clusters (operons) has an associated operator region and promoter. Each operator is however structurally similar and binds the same type of repressor.

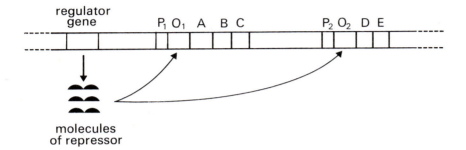

Figure 33

In the system shown in this diagram we assume five enzymes (a'-e') under the same controls, for which the structural genes (A-E) are in two clusters labelled 1 and 2 each with its own promoter (P1, P2) and operator region (O1, O2) alongside. O1, and O2 are the same in structure.

ITQ 9 (Objectives 8 and 9)

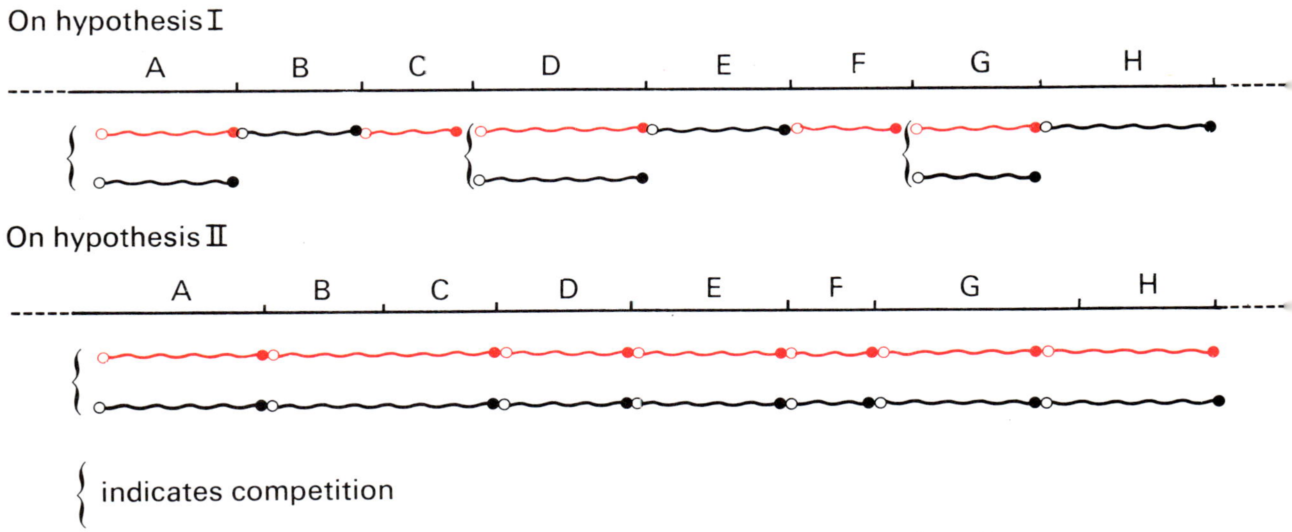

On hypothesis I

A B C D E F G H

On hypothesis II

A B C D E F G H

{ indicates competition

Figure 34

ITQ 10 (Objective 1) Yes. A particular mutant only goes through the changes characteristic of the stages before that at which it is blocked. This suggests that the changes characteristic of stages after the block cannot occur unless earlier ones are completed. It is also encouraging that the biochemical markers (Fig. 28) do appear to be intimately connected with particular stages and hence are useful as ways of characterizing those stages. (This does not however prove that they are necessary for those stages to occur.)

ITQ 11 (Objectives 10 and 13) There is a 40-minute gap between the completion of synthesis of the mRNAs and the appearance of the next sporulation event which is presumably dependent on translation of these mRNAs. It is unlikely that translation, as such, would take more than a few minutes. (Remember β-galactosidase starts to increase within 3 minutes of adding inducer to the cells and this time also includes transcription of the Z gene.) This implies that in the 40-minute gap other processes occur which then allow translation of the mRNAs formed earlier (as in the sea-urchin).

SAQ answers and comments

SAQ 1 (Objectives 1 and 2)
(a) False. It shows that all the cells have the same genetic information.
(b) True.
(c) False.
(d) True.
(e) True.
(f) False.

SAQ 2 (Objective 1)
(a) True.
(b) True.
(c) False. It determines the number of amino acids in the polypeptide chain.
(d) False.
(e) False. Protein degradation occurs.
(f) False.

SAQ 3 (Objectives 1 and 7)
(a) False. Repressor binds to an operator region.
(b) False.
(c) True.
(d) False. RNA polymerase catalyses RNA formation.
(e) True.
(f) False. A promoter is a region on the DNA where RNA polymerase binds.
(g) False. It controls the synthesis of several enzymes, not the structure.

SAQ 4 (Objective 7)
(i) (c)
(ii) (c)
(iii) (b)
(iv) (c)
(v) (b)
(vi) (b)

SAQ 5 (Objective 7)
(a) False.
(b) False. As protein turnover occurs anyway, some radioactive protein could be formed irrespective of an increased rate of synthesis or not.
(c) False. Inhibitors of protein synthesis would still be expected to reduce the normal rate of synthesis of E-ase which occurs as a part of protein turnover.
(d) True.
(e) False—although this might lead to all protein synthesis going up, not just that of E-ase.

SAQ 6 (Objectives 1 and 8)
(a) Yes.
(b) It would give the same results because, as all the cells are totipotent, they must all contain the same DNA.

SAQ 7 (Objectives 8 and 10) Not necessarily since if the two cells differ in their translational controls, then they may possibly translate the same types of mRNA with differing efficiencies and hence not produce the corresponding proteins equally well.

SAQ 8 (Objectives 1 and 13)
(a) False.
(b) True.
(c) False.
(d) False.
(e) False.

If you get any of the 'true or false' questions wrong and cannot see why from the answers, we suggest that you re-check the appropriate parts of the text.

S2—5   Genes and Development

Acknowledgements

Grateful acknowledgement is made to the following sources for illustrations used in these Units:

Fig. 1 R. Bulger; *Fig. 4* Academic Press Inc. and Dr F. C. Steward for F. C. Steward *et al., Current Topics in Developmental Biology*; *Fig. 24* The Clarendon Press, Oxford for Professor H. Harris, *Cell Fusion*; *Fig. 25* Syndication International for the Franklin Cartoon, *Daily Mirror*, 15 February 1965.